T0132517

Machine Learning, Deep Learning, Big Data, and Internet of Things for Healthcare

This book reviews the development of current technologies under the theme of the emerging concept of healthcare, specifically in terms of what makes healthcare more efficient and effective with the help of high-precision algorithms. The mechanism that drives it is machine learning, deep learning, big data, and Internet of Things (IoT)—the scientific field that gives machines the ability to learn without being strictly programmed. It has emerged together with big data technologies and high-performance computing to create new opportunities to unravel, quantify, and understand data-intensive processes in healthcare operational environments.

This book offers comprehensive coverage of the most essential topics, including:

- Introduction to e-monitoring for healthcare

- Case studies based on big data and healthcare

- Intelligent learning analytics in healthcare sectors using machine learning and IoT

- Identifying diseases and diagnosis using machine learning and IoT

- Deep learning architecture and framework for healthcare using IoT

- Knowledge discovery from big data of healthcare-related processing

- Big data and IoT in healthcare

- Role of IoT in sustainable healthcare

- A heterogeneous IoT-based application for remote monitoring of physiological and environmental parameters

Machine Learning, Deep Learning, Big Data, and Internet of Things for Healthcare

Edited by Govind Singh Patel, Seema Nayak, and Sunil Kumar Chaudhary

CRC Press
Taylor & Francis Group
Boca Raton London New York

CRC Press is an imprint of the
Taylor & Francis Group, an **informa** business

A CHAPMAN & HALL BOOK

First edition published 2023
by CRC Press
6000 Broken Sound Parkway NW, Suite 300, Boca Raton, FL 33487–2742

and by CRC Press
4 Park Square, Milton Park, Abingdon, Oxon, OX14 4RN

CRC Press is an imprint of Taylor & Francis Group, LLC

ISBN: 978-1-032-13082-8 (hbk)
ISBN: 978-1-032-13086-6 (pbk)
ISBN: 978-1-003-22759-5 (ebk)

DOI: 10.1201/9781003227595

Typeset in Minion
by Apex CoVantage, LLC

Contents

Preface

THE IMPORTANCE OF MACHINE learning, deep learning, big data and Internet of Things for healthcare is well known in almost all engineering fields. This book is structured to cover the key aspects of all these areas in the medical context.

This book uses plain and lucid language to explain the concepts of these subjects. It provides logical methods of explaining various complicated challenges and stepwise methods to explain important topics. Each chapter is well supported with necessary illustrations and practical examples. The chapters are arranged in sequence that permits each topic to build upon earlier studies. Care has been taken to make students comfortable in understanding the basic concepts of the subject.

This book not only covers the entire scope but also explains the philosophy of the subjects. This makes the understanding of the subjects clearer and more interesting. This book will be very useful not only to students but also to their teachers.

Acknowledgements

WE EXTEND OUR THANKS to Mr. Anil Bagane, Hon'ble Executive Director, and Dr. Sanjay A. Khot, Principal, SITCOE, Yadrav, Kolhapur, MH, India for giving us the opportunity to edit this book.

We owe a deep gratitude to Shri Mayank Agrawal Ji, Managing Director, Prof. (Dr.) M.K. Soni, Director General and Director, IIMT College of Engineering, Greater Noida, Uttar Pradesh, India, for providing us necessary facilities, valuable suggestions, and kind encouragement.

We are also thankful to Dr. Brijesh Kumar, Director, GCET, Greater Noida, for his support and guidance.

Dr. Govind Singh Patel
SITCOE, Yadrav, Kolhapur, Maharashtra, India

Dr. Seema Nayak
IIMT college of Engineering, Greater Noida, India

Dr. Sunil Kumar Chaudhary
GCET, Greater Noida, Uttar Pradesh, India

Editors

 **Dr. Govind Singh Patel** has a PhD in electronics and communication engineering from Thapar University, Patiala, India. He is a professor in electronics and telecommunication engineering at SITCOE, Yadrav, Kolhapur, MH, India. He has published more than 77 papers in national and international journals. He is a reviewer of many international journals, including for Springer and *Journal of Computer and Theoretical Nanoscience (JCTN)*.

 **Dr. Sunil Kumar Chaudhary** graduated (BE) from Nagpur University in 1996 in electronics, completed his post graduated (ME) in electrical engineering (power electronics, electric machines and electric drives) from MD University, Haryana, in 2006, and completed a PhD in electrical engineering, from Jamia Millia Islamia, New Delhi, in 2014. Dr. Chaudhary is a professor in the electrical engineering department of Galgotias College of Engineering (GCET), Greater Noida. His employment experience includes a tenure at the Lingaya Institute of Management and Technology (LIMAT) Faridabad, Haryana, Manav Rachna College of Engineering (MRCE), Faridabad, Haryana, Greater Noida Institute of Technology (GNIOT), Greater Noida, UP, and Gautam Buddha University (GBU), Greater Noida.

 **Prof. (Dr.) Seema Nayak** is a professor and head of the Department of Electronics and Communication Engineering at the IIMT College of Engineering, Greater Noida. Prof. (Dr.) Seema Nayak did her bachelor's in electronics and communications engineering at Maharashtra and received her PhD from MRIU, Faridabad. Her research area is digital signal processing. She has many publications in refereed journals and international conferences. She has filed two patents. She is also an editor and reviewer of many journals. She has 23 years of academic experience and has received grants for conferences, including FDP from AICTE, AKTU and ISTE. She is a member of the IEI IAENG Society. She has published research papers in many reputed journals.

Contributors

A.K. Awasthi
Department of Mathematics
Lovely Professional University
Phagwara, Punjab, India

Pranav Bakre
Department of Mechanical
    Engineering
Dr. Vishwanath Karad MIT World
    Peace University
Pune, Maharashtra, India

Bhawna Bakshi
State Transformation Manager,
PMSRI Lucknow
Uttar Pradesh, India

Vinod Bharat
Department of Electrical,
    Electronics and Communication
    Engineering
D. Y. Patil University
Ambi, Pune, India

Dr. Sunil Kumar Chaudhary
Department of Electrical
    Engineering
Galgotias University
Greater Noida, India

Arun Kumar Garov
Department of Mathematics
Lovely Professional University
Phagwara, Punjab, India

Pratik Gorade
Electronics and
    Telecommunication
    Engineering Department
Dr. Vishwanath Karad MIT World
    Peace University
Pune, Maharashtra, India

Dhiraj Gupta
Department of Electronics and
    Communications
Greater Noida Institute of
    Technology
Greater Noida, India

Ritom Gupta
Department of Electronics and
    Communications
Dr. Vishwanath Karad MIT World
    Peace University
Pune, Maharashtra, India

Vignesh Iyer
Electronics and
  Telecommunication
  Engineering Department
Dr. Vishwanath Karad MIT World
  Peace University
Pune, Maharashtra, India

Bharti Koul
Electrical Engineering
  Department
NIT Hamirpur
Himachal Pradesh, India

Sumit Koul
Mathematics and Scientific
  Computing
NIT Hamirpur
Himachal Pradesh, India

Kanhaiya Kumar
Department of Electrical,
  Electronics and Communication
  Engineering
Galgotias University
Greater Noida, India

Sanjeev Kumar
Department of Mathematics
Lovely Professional
  University
Phagwara, Punjab, India

Arundoy Lenka
Department of Electrical,
  Electronics and Communication
  Engineering
D. Y. Patil University
Ambi, Pune, India

P. Malathi
Electronics and
  Telecommunication
D. Y. Patil College of Engineering
Pune, Maharashtra, India

Shamla Mantri
School of Computer Science and
  Engineering
Dr. Vishwanath Karad MIT World
  Peace University
Pune, Maharashtra, India

Vishesh Kumar Mishra
Department of Electrical,
  Electronics and Communication
  Engineering
Galgotias University
Greater Noida, India

Baibaswata Mohapatra
Department of Electrical,
  Electronics and Communication
  Engineering
Galgotias College of Engineering
  and Technology,
Greater Noida, India

Ashish Mulajkar
School of Electrical and Electronics
  Engineering
Lovely Professional University
Phagwara, Punjab, India

Manoj Nayak
Mechanical Department, Manav
  Rachna International Institute
  of Research and Studies
Faridabad, India

Seema Nayak
Department of Robotics
  Engineering
ECE, IIMT College of
  Engineering
Greater Noida, India

Manish Pakhira
Department of Electrical,
  Electronics and Communication
  Engineering
The Neotia University
Kolkata, India

Govind Singh Patel
Department of Robotics
  Engineering
Sharad Institute of Technology
  College of Engineering
  (SITCOE)
Ichalkaranji, Maharashtra, India

Amrita Rai
Electronics and Communication
  Engineering, G.L. Bajaj,
  Institute of Technology and
  Management
Greater Noida, India

Kundankumar Rameshwar Saraf
Department of Electrical,
  Electronics and Communication
  Engineering
D. Y. Patil College of Engineering
Pune, Maharashtra, India

Snigdha Sharma
Department of Electrical,
  Electronics and Communication
  Engineering Galgotias University,
  Greater Noida, India

Rupali Shrivastava
Department of Electrical,
  Electronics and Communication
  Engineering
The Neotia University
Kolkata, India

Dinesh Singh
Department of Electrical,
  Electronics and Communication
  Engineering
Galgotias University
Greater Noida, India

Sanjeet K. Sinha
School of Electrical and Electronics
  Engineering
Lovely Professional University
Phagwara, Punjab, India

Lokesh Varshney
Department of Electrical,
  Electronics and Communication
  Engineering
Galgotias University
Greater Noida, India

CHAPTER 1

Introduction to E-Monitoring for Healthcare

Seema Nayak, Shamla Mantri, Manoj Nayak, and Amrita Rai

CONTENTS

1.1 INTRODUCTION

There has been a huge evolution of the embedded systems market due to the fast development of the connected devices. Internet of Things (IoT) is the system of embedded devices, software, sensors, and network connectivity that allows objects to collect and exchange data. IoT

DOI: 10.1201/9781003227595-1

permits objects to be sensed and controlled remotely across existing network infrastructure, creating opportunities for more direct integration between the physical world and computer-based systems, and resulting in better efficiency, accuracy, and financial benefits. Things or objects in the IoT sense can refer to a wide variety of devices, such as heart monitoring implants, biochip transponders on farm animals, electric clams in coastal waters, and automobiles with sensors. Radio-frequency identification (RFID) was seen as a prerequisite for the concept of IoT. If all objects and people in daily life were equipped with embedded devices and computers, then they could manage data and maintain record. In addition to RFID, things may be tagged through technologies as near-field communication, barcodes, QR codes, Bluetooth, and digital watermarking.

1.2 OVERVIEW OF IOT SYSTEM

IoT is a system with a mass of sensors, computers, and network connections designed into objects (embedded systems) to enable centralized control. It has a huge potential, and its efficiency creates significant economic value. It provides an opportunity to update and configure the functionality of devices through the cloud platform.

Wearable health monitoring systems are drawing more and more attention from the research community. Wearable technology, also known as wearables, fashionable technology, wearable devices, fashion electronics, or tech togs, are modern smart electronic devices equipped with a microcontroller and sensor that can be worn on the body as accessories. They mostly involve electronics, software, and IoT. The data of the system is stored either in a device or on cloud platforms where data can be accessed and monitored without human intervention.

IoT offers several advantages in the field of healthcare systems and medical equipment. The first utilization of IoT in this field began during the COVID-19 pandemic, when doctors were not physically available in the hospital. Patient contacted doctors through some IoT-specific application and booked their appointments using their mobile phones. IoT can help healthcare systems from all perspectives, including by helping patients and their family keep records and monitor health conditions, helping doctors contact patients and their family members, and assisting hospitals and insurance companies [1], [2].

A lot of development and exploration in the area of healthcare services is based on the wireless sensor network (WSN) [3], [4], which is an initial effort on the basis of research in the healthcare sector in IoT. However, the emerging low-power wireless IPv6-based personal area networks can be used to trade the current trend. There is need for critical analysis when WSNs becomes a central part of the internet [5]–[7]. The main issues in healthcare systems are related to diagnosis, treatment, health professionals, and policies and public health. The particular healthcare system contacts an organization to identify and deliver healthcare services to meet the needs of smart healthcare. The smart healthcare system based upon IoT has e-health and smart devices which are used for upgradation and future smart healthcare technologies [8], [9]. The World Health Organization (WHO) has publicized various plans to set up the Disease Intelligence Unit, which will operate independently. After the introduction of smart disease surveillance in the healthcare industry, the process of surveillance will speed up to reach the highest goals of accuracy and real-time databases [3].

1.3 ADVANTAGES OF IOT IN HEALTHCARE SYSTEMS

- The innovative healthcare system has made it possible for doctors to practice remote monitoring, especially for chronic disease management and care for elderly patients [8], [10].

- Advantages of healthcare monitoring systems are that they use less energy and extend the communication coverage because of the use of WSNs [11].

- Most developing countries are facing problems like lack of updated technology and minimum availability of smart devices and smart objects, which is a universal need of smart healthcare systems [3].

- Many developing devices like heart monitoring devices use wireless sensors and smartphones, which detect the threatening arrhythmias and alert the patient when they reach certain threshold values [10].

- When we apply IoT to personalized healthcare in smart homes, it gives services and technological approach [11].

- Today, smart healthcare systems are used for disease observation. These are mainly categorized into smart IoT devices and smart backbone devices. Generally, they operate the functioning of cloud computing and the main servers at the hospitals.

- IoT is used to develop architecture for smart health systems using sensors like temperature, barometric pressure, and ECG sensors. It facilitates remote monitoring and management of emergency situations [4].

Throughout the COVID-19 pandemic, there has been a need to study various applications of IoT-enabled healthcare systems. Cutting-edge information technologies have opened a new door to innovation in our everyday lives. IoT is a developing technology that provides enhancement and good solutions in the medical field, like proper medical record keeping, sampling, integration of devices, and tracking causes of diseases. Sensor-based IoT technology provides an excellent way to reduce the risk of surgery during complicated cases and is helpful in COVID-19-type pandemics. In the medical field, IoT's focus is to perform the treatment of different COVID-19 cases accurately. Using new technology to minimize risks and increase overall performance also make a surgeon's job easier. This technology opens up many unique healthcare opportunities for medical students, who can be trained for disease detection and for the future course of action. Use of IoT can help to resolve different medical challenges like speed, price, and complexity. It has also improved the overall performance of healthcare system during the COVID-19 pandemic.

Future smart healthcare systems, also referred to as the Internet of Medical Things (IoMT), will combine a plethora of wireless devices and applications that use wireless communication technologies to enable the exchange of healthcare data. Smart healthcare requires sufficient bandwidth, consistent and safe communication links, energy-efficient operations, and quality of service (QoS) support. The integration of IoT solutions and healthcare systems can significantly raise intelligence, flexibility, and interoperability. This chapter provides a wide survey on emerging IoT technologies suitable for smart healthcare applications.

1.4 IOT IN HEALTHCARE

Since the number of patients of diverse diseases is increasing, the necessity of monitoring patients' health outside of the hospital has increased. To fulfill this requirement, many system prototypes and commercial products have been developed. The wearable health monitoring system is a collection of wearable sensors that are able to monitor health-related issues. Wearable systems are small electronics arrangements that anybody can wear easily, as shown in Figure 1.1.

Quickly increasing aging populations and associated challenges in health- and social care raise the costs of healthcare sky high. There has been a need to monitor patients from a remote location. Take, for example, Alzheimer's disease, which slowly destroys brain cells

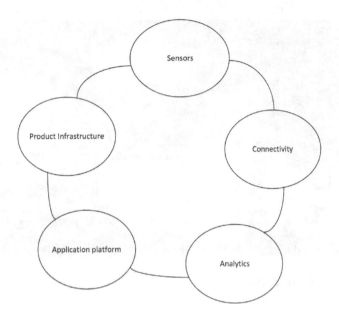

FIGURE 1.1   IoT in healthcare

and causes elderly persons to forget things and past events. Many people spend more than \$1,285 per year caring for someone with Alzheimer's. Many elderly people die due to this disease because they forget to take medicine on time, and sometimes they fall down accidently. Researchers have developed a portable, easy-to-use, and cost-effective system that uses a panic button. If a person needs any help then they press the button, which sends an alert message to their caretaker. This innovation helps such patients.

1.5 SMART DEVICES

Since the introduction of IoT in the healthcare system, as shown in Figure 1.2, researchers have accepted that IoT is one of the most sophisticated technologies with the highest prospects of innovation. Smart devices are used to treat issues such as chronic disease management, personal health, and fitness management.[12]

1.6 CORONA PROTECTION WATCH

IoT-enabled Corona protection watches (Figure 1.3) can protect people from contracting COVID-19. It follows social distancing, maintains record, is easy to use, and is easy to carry. It can be easily charged.

FIGURE 1.2 Smart devices.

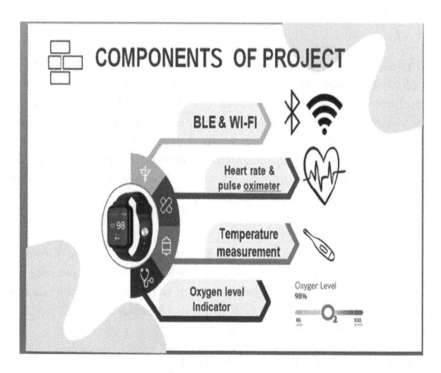

FIGURE 1.3 Corona protection watch.

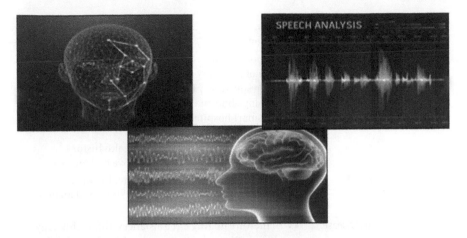

FIGURE 1.4   Approach for depression analysis.

Depression analysis can be done using signal processing, and using IoT, results can be sent to a counsellor, as shown in Figure 1.4.

1.7 IOT-ENABLED DEVICES FOR HEALTHCARE MONITORING

IoT-enabled devices offer a number of novel prospects for healthcare professionals to monitor patients and for patients to monitor themselves. By extension, wearable IoT devices provide a collection of benefits and challenges for healthcare providers and their patients alike.

In IoT-based services, a lot of devotion is required for IoT healthcare applications. These IoT applications require more services since they are used by patients, as shown in Table 1.1.

TABLE 1.1   IoT Applications [13]–[15]

SN	IOT Applications	Description
1	COVID-19 Patient Treatment	• Using GPS, IoT-enabled devices are used for the treatment of the COVID-19 patients • Medical devices like wheelchairs, nebulizers, scales, pumps, and others are used for observing • Monitoring and controlling environmental conditions like temperature, humidity, etc. as per the requirement of patients

(Continued)

TABLE 1.1 (Continued)

SN	IOT Applications	Description
2	Smart Hospital	• IoT provides lists of smart hospitals in nearby locations • Using software, proper information regarding ongoing abnormalities of the patient can be provided • The smart hospital gives all relevant information and reduces waiting time • It provides analysis of patient health history • Using data analysis, helps improve patient care
3	Data Storage of Patients	• For better treatment in the future, IoT-enabled devices can transmit, efficiently store, and analyse patient data
4	Alerts about Disease	• In life-threatening circumstances, this technology alerts patients about disease with real-time tracking • Provides notification to people via linked devices • Provides reports about the condition of human health
5	Proper Medication	• IoT-enabled devices monitor proper medication to the body and suggests proper diet for the patient • Traces and monitors the condition of the patient in daily life
6	Proper Facilities	• IoT provides many facilities in healthcare during the pandemic • Exchanges information for effective healthcare services
7	Checking of Glucose Levels	• Monitors the level and flow of glucose • Automatically adjusts insulin amounts
8	Assistance in Remote Areas	• In urban areas, IoT can help remotely located patients, providing doctor contact information using mobile phone applications • IoT improves patient care and digitization in the hospital
9	Detection of Asthma/ Heart Attacks	• Using IoT quickly predicts the symptoms of asthma/ heart attack before an attack can occur • Immediately sends notification about the attack and related information
10	Reminders about Medication Time	• Reminds elderly patients to take their medication • Reminds the patient to take medicine at the prescribed time
11	Emergencies	• In case of an emergency, IoT analyses the distance and profile before reaching the patient to nearest hospital • Improves emergency care and also reduces family loss
12	Availability of Smart Beds	• IoT applications give availability of beds in hospitals • IoT enabled smart beds can automatically adjust pressure and support to the patient as per requirement
13	Robotic Surgery	• Inserting small IoT-enabled robots inside the human body, surgeons can perform complex surgeries

1.8 CHALLENGES AND CURRENT ISSUES IN IOT-BASED HEALTHCARE SYSTEMS

Today, IoT is emerging technology that is growing exponentially. It is likely that the healthcare sector will fully embrace IoT technology and that it will flourish as the result of modern IoT-healthcare applications and devices. These healthcare applications and devices using IoT are expected to contain important information, including personal healthcare records.

Many researchers have focused on implementing and designing different types of innovative IoT healthcare frameworks and on resolving numerous architectural complications related to these frameworks. However, there are still many open research problems and challenges that need to be satisfactorily addressed.

In some cases, long-term monitoring of patient health is required. For these situations, continuous registration and monitoring is needed. Thus, design of low-cost IoT-enabled prototypes must be considered. Most of the medical devices have less integrated memory and they can be activated with an integrated operating system.

Healthcare organizations and hospitals often deal with many patients, with multiple support personnel performing various tasks. Accurate identification of patients and staff is required to achieve proper data management. Proper innovation and classification of all IoT devices on a healthcare provider's network helps guard against this risk. Once IoT device networks are properly identified, classified, regulated, and secured, managers can track device behaviour to identify anomalies and perform risk assessments.

1.9 CONCLUSION

The features of a smart healthcare system has been enhanced through the latest technology. The smart healthcare system has reduced complication and complexity with the introduction of IoT. The IoMT is the collection of medical devices connected through networks. Many healthcare professionals use IoMT applications to increase the efficiency of treatments, regulate diseases, minimize errors, and decrease system costs. Recently, wearable devices in healthcare services have steadily increased, and the paradigm has changed from being centred on diagnostic therapy to prevent pre-diagnosis through exercise involvement and lifestyle changes.

Security is a precarious requirement; both legal regulations and user concerns have to be taken into account. IoT could address the developing needs of patients and healthcare providers at a reasonable cost by using today's healthcare systems and existing structures.

Hence, various medical devices, sensors, diagnostics, and imaging devices can be seen as intelligent devices and become a central part of IoT. These services reduce costs, increase quality of life, and enrich user experience. IoT in healthcare has the potential to reduce device downtime through distant delivery.

1.10 FUTURE SCOPE

In the future, IoT will observe vital signs of the patient in a real-time situation. There will be a major improvement in healthcare practice using the latest technologies, and doctors will have to use them. IoT is a sophisticated emerging technology with wide applications in providing precise medical care that opens up an effective way to analyse valuable data, information, and testing. The future of IoT in healthcare has many applications in managing inventories used in the medical field and the medical supply chain for getting the right item at the right time and location. Such innovation in the information system will enable smart healthcare service in the medical 4.0 environment.

REFERENCES

[1] Takabayashi K., Tanaka H., Sakakibara K. (2018), "Integrated Performance Evaluation of the Smart Body Area Networks Physical Layer for Future Medical and Healthcare IoT", *Sensors.* 19(1). doi: 10.3390/s19010030. PII: E30.

[2] Arun M., Baraneetharan E., Kanchana A., Prabu S. (2020), "Detection and Monitoring of the Asymptotic COVID-19 Patients Using IoT Devices and Sensors", *International Journal of Pervasive Computing and Communications, ID: covidwho-862493.*

[3] Mathew A., Farha Amreen S.A, Pooja H.N, Verma A. (2015), "Smart Disease Surveillance Based on Internet of Things (IoT)".

[4] Sreekanth K.U., Nitha K.P. (2016), "A Study on Health Care in Internet of Things". 4(3):44–47.

[5] Riazul Islam S.M., M.D., Kabir H., Kwak D., Kwak K.-S., Hossain M. (2015), "The Internet of Things for Health Care: A Comprehensive Survey", 10.1109/ACCESS.2015.2437951.

[6] Natarajan K., Prasath B., Kokila P. (2016), "Smart Health Care System Using Internet of Things", *Journal of Network Communications and Emerging Technologies.* 6(3).

[7] Miorandi D., Sicarib S., De Pellegrinia F., Chlamtac I. (2012), "Internet of Things: Vision, Applications and Research Challenges", *Ad Hoc Networks.* 10(7):1497–1516.

[8] Khanna A., Misra P. (2014), "The Internet of Things for Medical Devices Prospects, Challenges and the Way Forward".

[9] Ozdemir V. OMICS. (2019), "The Big Picture on the 'AI Turn' for Digital Health: The Internet of Things and Cyber-Physical Systems". 308–331. doi:10.1089/omi.2019.0069.

[10] Padwal S.C., Kurde S.V. (2016), "Long-Term Environment Monitoring for IoT Applications Using Wireless Sensor Network".

[11] Yu L., Lu Y., Zhu X.J. (2012), "Smart Hospital Based on Internet of Things".

[12] Warren S., Richard L. (1999), "Designing Smart Health Care Technology into the Home of the Future".

[13] Ndibanje B., Lee H.J., Lee S.G. (2014), "Security Analysis and Improvements of Authentication and Access Control in the Internet of Things", *Sensors*. 14(8):14786–14805.

[14] Lomotey R.K., Pry J., Sriramoju S. (2017), "Wearable IoT Data Stream Traceability in a Distributed Health Information System", *Pervasive and Mobile Computing*. 40(C):692–707.

[15] Lin H., Garg S., Hu J., Wang X., Piran M.J., Hossain M.S. (2020), "Privacy-Enhanced Data Fusion for COVID-19 Applications in Intelligent Internet of Medical Things", *IEEE Internet of Things Journal*. (99):1. doi:10.1109/JIOT.2020.3033129.

CHAPTER 2

Case Study-Based Big Data and IoT in Healthcare

Arun Kumar Garov and A.K. Awasthi

CONTENTS

DOI: 10.1201/9781003227595-2

2.1 INTRODUCTION

Big data is a group of data that is very large and continuously gets bigger and bigger over time. Mobile devices, social media, and the many sensors being used today generate a lot of data at a very rapid rate. This data is growing very fast all over the world. With the help of new technology and the base of the models, the generation of data is also happening rapidly. Internet of Things (IoT) is also contributing to the generation of data.

If someone working on Big data technology, it does not mean that size of storage is big. It is a technology used for huge amount of data analysis. Big data is known/identified by the following attributes for proper understanding: volume, velocity, and variety. These are also called the three Vs.

The varieties of big data are shown in Figure 2.1.

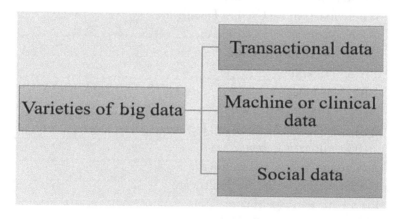

FIGURE 2.1   Varieties of big data.

1. **Transactional data:** This data includes invoices, market sales and purchases, storage records, payment orders, cost data, and other dates included in forms mean time series. Transactional data is used in healthcare for payers and providers.

2. **Machine or clinical data:** This data can be collected by sensors, industrial equipment, weblogs, etc.

3. **Social data:** Social data can be collected by social sites, such as Facebook, Instagram, YouTube, and Twitter. In healthcare, this type of data helps providers understand the behaviours or sentiments of the patient.

2.2 WHERE CAN WE COLLECT BIG DATA?

Many industries, companies, and agencies, academicians, etc. collect data in a big form, which helps in analyses for future forecasting. The following list includes sectors using big data:

1. Transportation

2. Banking

3. Security

4. Education

5. Entertainment

6. Media

7. Insurance

8. Government

9. Wholesale and retail trade

10. Electricity

11. Healthcare

12. Agricultural

Big data is used by companies to improve and develop operational efficiency, as well as to provide better, more convenient service to the customer to generate increased profits and growth in market share.

2.2.1 Examples

In the health sector, medical researchers use big data to identify the symptoms of diseases and identify harmful factors. Similarly, big data is also used to solve medical and clinical situations of patients. Electronic health records and the internet, especially media sites, provide government agencies and healthcare organizations with the effects of diseases and their symptoms, with details being updated from time to time.

2.3 BIG DATA IN THE HEALTHCARE SECTOR

In healthcare, data analysis is making a big change in this current era. This is helpful for the newcomers and patient treatment. Big data is useful for finding new ways to treat diseases. It can benefit patients and providers for healthcare in different ways, including the following:

1. **Effect on Patients:** Big data is used to understand the personal and community trends of the patient and create a plan for useful treatment or forecast of patient risk.

2. **Staff and Operating Systems:** Big data can be used to relieve overloaded of staff and the requirement of new staff members in healthcare and shift-wise arrangement of staff members. It is also used to forecast the long-term treatment of patients coming to the healthcare centre.

3. **Development of Product:** Big data can help in researching a new product, new therapy, and medicine for healthcare.

4. **Strategic Planning:** Healthcare analytics can be used to identify harmful diseases and in plans for solving problem.

Sources of big data in the healthcare sector are shown in Figure 2.2.

2.4 ROLE OF BIG DATA IN HEALTHCARE

Big data has also contributed to the change of the world. Over time, many industries and sectors have been transformed with big data. Patient get better and productive information by this health care system. Big data provide

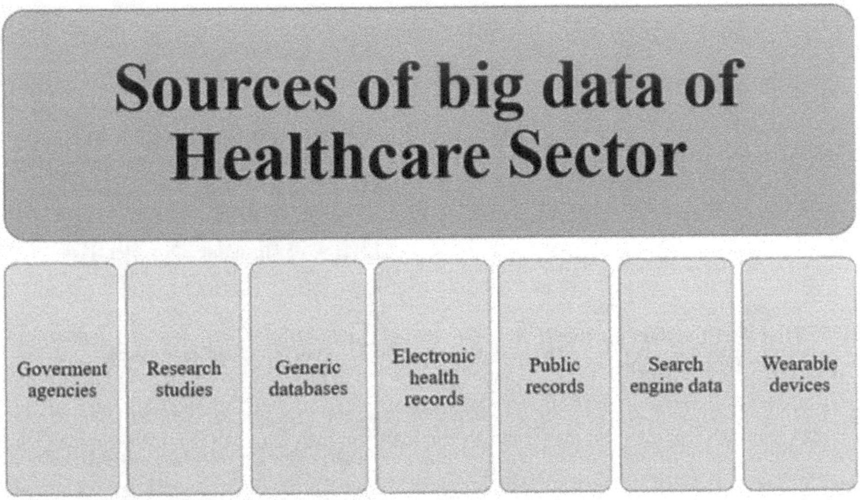

Sources of big data of Healthcare Sector

Goverment agencies	Research studies	Generic databases	Electronic health records	Public records	Search engine data	Wearable devices

FIGURE 2.2   Sources of big data in the healthcare sector.

investigation in positive way to live better life to patients. Different types of data analytics tools and methodologies are used for the patient in the form of care healthcare, intervention activities for health management, etc., which improves patient experience and care and decreases cost. It makes easy to understand to anything, similarly big data work as an image to make better decisions in healthcare.

In December 2019, COVID-19 was introduced to the world. Its effects are very harmful to the body and is highly transmissible. With the number of cases increasing, data is also increased and converted into big data with different parameters, i.e., positive persons, deaths, negative persons, symptoms, etc. the healthcare industry created massive amounts of information about COVID-19. COVID-19 data is currently an example of big data.

2.5 BIG DATA SOURCES FOR HEALTHCARE

Sources of big data for healthcare come in three main types: structured data, semi-structured data, and unstructured data. These three types of big data sources for healthcare are shown in Figure 2.3.

In the past, management and analysis of data was difficult and costly for all sectors, including healthcare. Big data analytics use information to help understand any patient condition. It is now easy to check patient

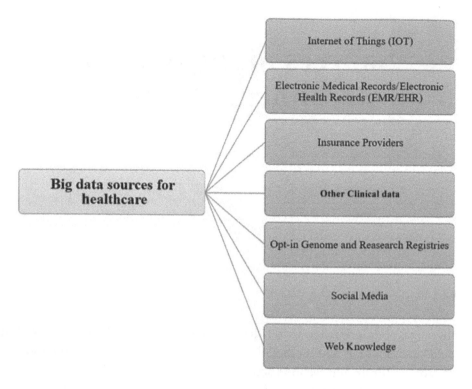

FIGURE 2.3   Big data sources for healthcare.

information because every healthcare sector has systems that store patient and other kinds of data.

Big data analytics have six types of analytics for every business. In the field of healthcare big data analysis is performed in three types of analytics: descriptive analytics, predictive analytics, and prescriptive analytics.

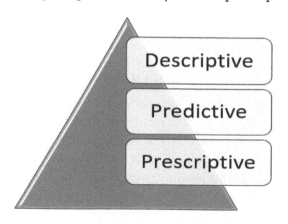

2.6 INTRODUCTION TO INTERNET OF THINGS (IOT)

Everything is possible in the world by Human. Humans invented many things in every sector for getting more. IoT is also an example of that. IoT is a platform, system, etc., but is also a collection. What type of collection? It is a collection of data connected to human activities with devices, sensors, software, etc. where data can be shared and stored. How is data generated? If someone is working on the internet and their history is being saved automatically, this search history is a part of data, and this data is shared. In the same way, IoT connects to the digital and physical universes which makes it seem smart and attractive. The following are types of Internet of Things:

Types of Internet of Things (IoT)

1. Consumer IoT

2. Commercial IoT

3. Military Things (IoMT)

4. Industrial IoT

5. Infrastructure IoT

The second type of IoT is commercial, which is used in healthcare. Take some impact on healthcare.

2.7 INTRODUCTION OF HEALTHCARE

Health care is a department to motivate to the member of health-related issues and this department member is to help everyone who is effected by it mentally and physically. For the patient, healthcare is separated into primary care, specialty care, emergency care, urgent care, long-term care, hospice care, and mental healthcare.

During the COVID-19 pandemic, the biggest market was hospitals and clinics, as well as the health sector more generally. The reason is that everywhere was locked down and no one could go anywhere. Only those persons could go outside who were facing health problems, otherwise there was no reason to leave the home. Everything was provided near home. When the situation settles down then every sector will open, but today a big market is the health sector. Healthcare is a big platform. Problem of people such as physical and mental can be resolved using such technology. Some areas using IoT in the healthcare market include clinics, hospitals, research centres/labs, diagnostic labs, government institutes, defence institutes, research organizations, etc.

In the future, IoT will work as a game changer for healthcare because it has the capacity to update techniques for treatments, reduce costs, and improve systems of care and drug development.

2.8 WHERE IOT CAN BE USED IN HEALTHCARE

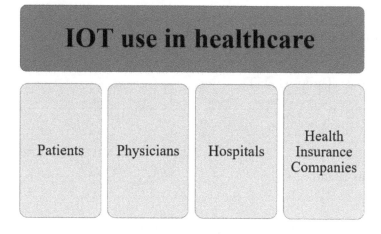

2.9 IOT APPLICATION IN HEALTHCARE

IoT is used in every field for different purposes, but it is useful and interesting for everyone. When discussing IoT in healthcare, it is important to note that it helps in several ways, including:

1. Reducing wait times in the emergency room
2. Tracking for patients, inventory, and staff

3. Developing drug management

4. Ensuring availability of critical hardware

2.10 DEVICES IN HEALTHCARE IOT

IoT based technology is used everywhere in our daily life. Which is helpful for patient to think positive for long life and improve their lifestyle. These devices include the following:

2.10.1 Hearables

Some people cannot hear from birth, and some people can simply not hear due to an accident. Some devices have been developed by IoT to understand and solve their hearing problems. But whether every device is capable or not is also important. Therefore, equipment has been developed according to every problem and they are also updated from time to time. Hairless is a device that is available in today's time for non-hearing people. Which helps them to listen and keeps them connected with the people and the world. These devices are also available with Bluetooth, examples of which are Doppler Labs, etc.

2.10.2 Inhalers

Asthma is a respiratory disease in which the patient has difficulty breathing in the presence of airborne particles. If the patient does not take care of the convenience, then suddenly there may be trouble, so it becomes necessary for the patient to have an inhaler. In today's era, the inhalers connected to Internet of Things are an the inhalers is connected to patient with IoT to avoid the problem, which was polluted due to environment around the patient.

2.10.3 Ingestible Sensors

Electrical science made a pill-size sensor device. This device monitors the quantity and regularity of medicine intake and is useful for diabetes patient. It warns of disease and indicates the effectiveness of medicines.

2.10.4 Glucose Monitoring Devices

This device is used for checking levels of glucose and helps to advise in difficulties. Manually it is very difficult, but it is possible to measure glucose level automatically by IoT.

This device is small in size so there is no problem in monitoring by the patient. There is also no need to charge often, as it uses a small amount of power.

2.10.5 Smartwatches

There are smartwatches available on the market that can be purchased at electronics stores. These devices have sensors and internet connection facilities. Some devices have some other features, such as monitors for heart rate, speech treatment, diabetes control, etc. The iWatch series 4 is a device available on the market for monitoring your heart rate.

Similarly, Moodables, ingestible sensors, Arduino, etc. are also useable devices in the healthcare sector.

2.11 ADVANTAGES OF IOT IN HEALTHCARE

The advantages of IoT in healthcare extend not only to the patient but also to doctors, researchers, nursing staff, etc.

- **Efficient Use of Resources:** If we want to increase the efficiency of resources then we need to understand the work of each device in terms of capacity. Along with this, natural resources can also be taken into account.

- **Reduced Human Efforts:** Everyone uses devices in daily life, which increases the speed of work, but is reducing human effort day by day good? Think about this.

- **Time Duration:** If anyone uses such device then work will be finished fast. This means that these devices could save people time.

Similarly, we can monitor or manage many more such activity like research and other managerial activity etc. in health care.

2.12 DISADVANTAGES OF IOT IN HEALTHCARE

Advantages are often accompanied by disadvantages. Just as a coin has two sides, that which is beneficial to us also can cause harm. The following lists the disadvantages of IoT in healthcare areas:

- **Security and Privacy:** While use of connected devices in healthcare for saving time and fast working process benefits users, user fears persist about security and privacy. Because many different kinds of hackers are in the world, any kind of hacking and infraction may be in the monitoring system of the healthcare. This is a big problem for healthcare IoT, which can have serious consequences.

- **Unification:** Production of devices by different manufacturers is may be reason of not a good work. Because, by regarding standards and protocols of IoT is no any consensus. Uniformity lack is not required, if it is then it make a full stop on unification, so its probable effectivity bounded.

- **Risk of Failure:** Many kinds of risk effect healthcare operations, such as power failures, which can affect sensors and equipment, cause hardware crashes or bugs, etc. Interrupting scheduled software updating is also dangerous.

- **Cost:** In the past, IoT always promised to reduce the cost of healthcare, but there has been no change. Its implementation cost is very high in hospitals and staff training.

2.13 IOT'S FUTURE IN HEALTHCARE

In past and present, the IoT device market is increased, mean is IoT devices are using much more in the market, but can't say that it is being used in only one sector. It is being used in every sector and is also being more used in the healthcare sector. If we think for the future, then IoT needs to improve security, high-speed communication, privacy, compliances, etc.

Many IoT devices are currently used for the improvement and development of the healthcare sector, and this will continue in the future.

2.14 WORKING TOGETHER: BIG DATA AND IOT

Big data and IoT work together in a relationship of data to source. Actually, when IoT will more grow then requirement will increased in market and data also increased automatically. Companies are use sensor device for collect the data and transmit. Big data are store in both forms as structured and unstructured. Analytics methods are used to understand data in easy form/other form such as; graphically, reports, etc. Devices are also use for future matrices. Both are corresponding to each other, requirement of even-more development for future.

2.15 MIXED IMPACTS OF BIG DATA AND IOT

Mix impacts of both big data and IoT help to take efficient decisions for future in every sector companies. Let see some effects on companies:

1. For data checking

2. For hidden relations

3. For hidden data frameworks

4. For unfolding new information

5. For benefits in the transport industry

6. For helping to increase business ROI

2.16 BIG DATA FOR PREDICTIVE ANALYTICS

In Section 2.5, we discussed big data analytics types. Predictive analysis is a type of big data analysis. Predictive analysis (forecasting analysis) recognises meaningful patterns of big data. After the recognise the patterns forecast the future. Any data of any types are capable for forecasting. So, here we discuss predictive analytics in healthcare and says that forecasting analytic in healthcare.

2.17 HEALTHCARE FORECASTING

Forecasting is a process about the future, which depends on past and present situations. To understand future data mathematically or graphically, one needs past and present data. The experiment of models applies to data for the forecasted future. but it will fully correct it is not clear. As Neils Bohr said, "Prediction is very difficult, especially about the future" [3].

Everyone wants to know about the future. How will our future? It is a big problem in every person's life, and involves many fields such as business, industry, economics, medicine, finance, etc [1, 2]. Forecasting can be applied in the form of one day, one month, one year, or many years; that is, forecasting can be performed in three timeframes: short-, medium-, and long term.

Mostly time series data are used in forecasting problems. Figure 2.4 shows the time series plot of BSE Healthcare data from January 2008 to July 2021 [3].

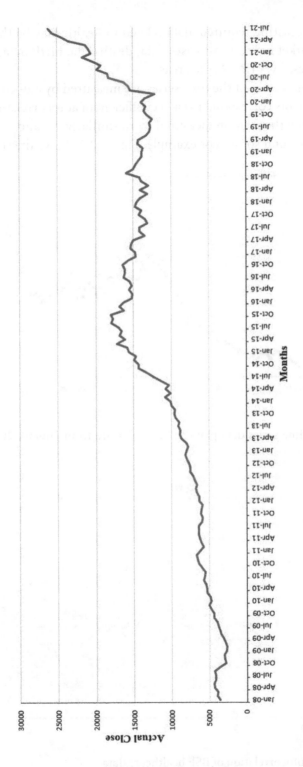

FIGURE 2.4 Time series plot of BSE Healthcare data from Jan 2008 to July 2021.

For the forecasting required, data related to health may be the health-care stock market index, heart cases data, death data, birth data, effective by any diseases, COVID-19 data areas.

Lagging data values of the time series are measured by autocorrelation. Lagging data plots correspond to each coefficient of autocorrelation. For y_t and y_{t-1} between the relation measured by r_1, similarly y_{t-1} and y_{t-2} relation measured by r_2, and so on. For example, Figure 2.5 shows the time series

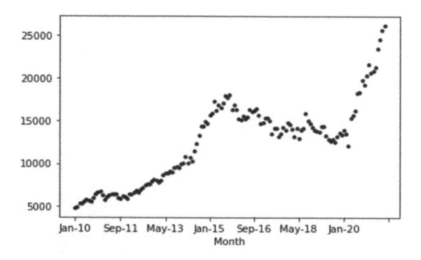

FIGURE 2.5   Time series data plot of BSE healthcare from January 2010 to July 2020.

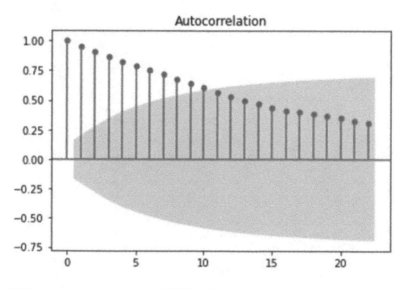

FIGURE 2.6   Autocorrelation of BSE healthcare data.

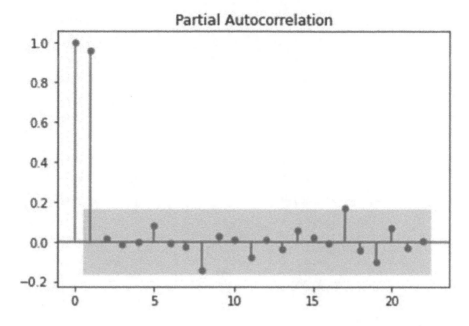

FIGURE 2.7　Partial autocorrelation of BSE healthcare data.

of data. The autocorrelation and partial autocorrelation of this data are represented in Figures 2.6 and 2.7.

For time series data three types of modelling are used: (1) Autoregressive (AR), (2) Moving average (MA), and (3) Autoregressive integrated moving average (ARIMA). Here we will discuss moving averages.

2.18 MOVING AVERAGES

Moving averages are a statistical method used for long-term trends of forecasting [4]. Estimation of a trend at time t is found by averaging a time series value for k period of t. Formulation of moving average of m order is,

$$T_t = \frac{1}{m} \sum_{k}^{i=-k} y_{t+i}, \ m = 2k+1$$

It is also called the moving average of order m (m-MA). if value of k is 2 then order is m = 2k + 1 = 5 (five), similarly, k = 3, m = 7, and k = 4, m = 9, etc. Figure 2.8 shows the order-changing effects of moving averages for BSE healthcare market index data.

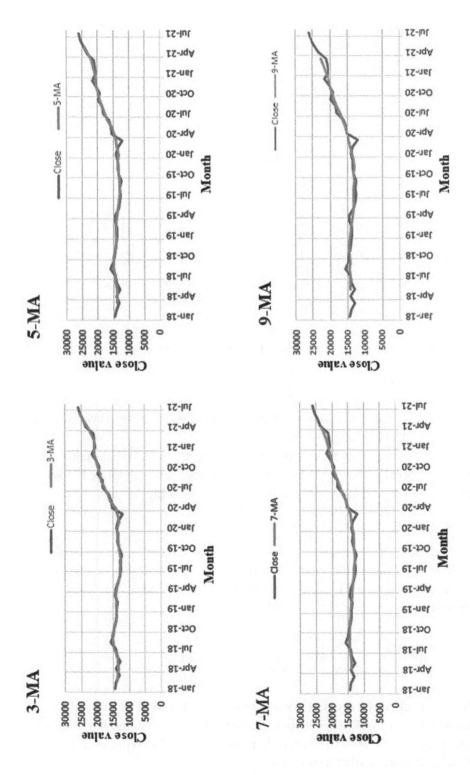

FIGURE 2.8   Different types of order moving averages applied to the BSE healthcare market.

2.18.1 Example of Forecasting

For the healthcare forecasting analytics, we have taken the BSE Healthcare market index month wise historical data from January 2018–July 2021. In this process January 2018–December 2019 using for training purposes and January 2021–July 2021 for testing purposes. Figure 2.9 shows the BSE market index from January 2018–July 2021.

Define the universal set S, $S = \left[actual_{min} - D_1, actual_{max} + D_2 \right]$, where, D_1, D_2 are proper positive real values [2, 5]. So, the universal set is S=[12145, 26245] and the length of an interval is 100 according to quantity based historical data then intervals are $s_1, s_2, s_3, \ldots, s_{141}$. These intervals are framed into fuzzy set β_τ $(\tau = 1,2,3,\ldots,141)$. Fuzzy set β_τ in universe S is defined as follows;

$$\beta_\tau = \frac{f_{\beta_\tau}(s_1)}{s_1} + \frac{f_{\beta_\tau}(s_2)}{s_2} + \frac{f_{\beta_\tau}(s_3)}{s_3} + \ldots + \frac{f_{\beta_\tau}(s_m)}{s_m}$$

Where f_{β_τ} is relationship function of the fuzzy set β_τ and $f_{\beta_\tau}(s_j)$ represents the degree of connection of (s_j) belonging to the fuzzy set β_τ.

$$f_{\beta_\tau}(s_j) \in [0,1], \quad 1 \le j \le m$$

Suppose Y(t), t = . . .,−2,−1,0,1,2,3, . . . be the universe of discourse and be a subset of R. Let $f_\tau(t)$, where $\tau = 1,2,3,\ldots$ be the fuzzy set, which is defined in

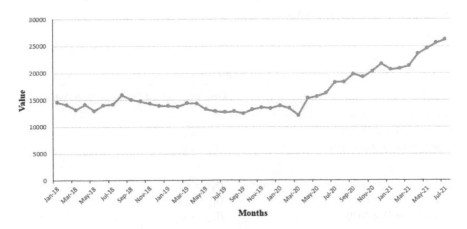

FIGURE 2.9   BSE healthcare market index, January 2018–July 2021. (Source: www.bseindia.com/Indices/IndexArchiveData.html.)

the universe of discourse Y(t) and let F(t) be a set of $f_\tau(t)$ then the set F(t) is called a Fuzzy time series of Y(t).

Definition: [2, 6, 7] If a fuzzy relationship R(t, t+1) exists, such that $F(t+1) = F(t) \circ R(t, t+1)$ where "$\circ$" denote the max- min composition operator, then F(t+1) is called the caused by F(t) and quantity based fuzzy relationship is,

$$F(t) \rightarrow F(t+1)$$

Both F(t) and F(t+1) are fuzzy sets.

Suppose F(t) and F(t+1) are β_τ and β_j, respectively. The relation between these two fuzzy sets is represented in the form "$\beta_\tau \rightarrow \beta_j$". β_τ and β_j are left-hand and right-hand sides of a fuzzy logical relationship. A quantity based fuzzy logical relationship makes a quantity based fuzzy logical relationship group (FLRGs) as follows;

$$\beta_\tau \rightarrow \beta_{j1}, \beta_{j2}, ..., \beta_{jm}$$

A fuzzy set generates a relationship of monthly data and that relationship is called a fuzzy logical relationship. A fuzzy logical relationship makes a group, and these groups are called quantity based fuzzy logical relationship groups.

At the relationship groups assigned weights for fuzzy set $\beta_1, \beta_2, \beta_3, ..., \beta_{141}$ is $w_1, w_2, w_3, ..., w_{141}$, respectively.

TABLE 2.1   Quantity-Based Fuzzy Logic Relationship January–Dec 2020

Month	Quantity-Based Fuzzy Logic Relationship
Jan 2018 → Feb 2018	$\beta_{25} \rightarrow \beta_{20}$
Feb 2018 → March 2018	$\beta_{20} \rightarrow \beta_{11}$
March 2018 → April 2018	$\beta_{11} \rightarrow \beta_{21}$
⋮	⋮
Sep 2020 → Oct 2020	$\beta_4 \rightarrow \beta_{11}$
Oct 2020 → Nov 2020	$\beta_{11} \rightarrow \beta_{15}$
Nov 2020 → Dec 2020	$\beta_{15} \rightarrow \beta_{13}$

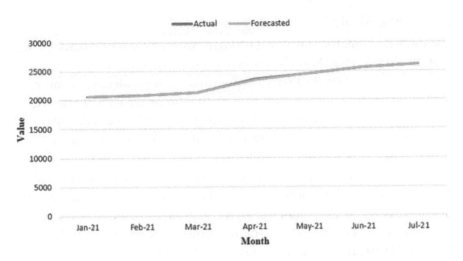

FIGURE 2.10   Actual and forecasted value of BSE Healthcare, January–July 2021.

Result-calculation: The final forecasted value F(t) is equal to the product of defuzzified matrix and transpose of the weight matrix:

$$F(t) = M(t) \times W(t)^T = \left[m_{j1}, m_{j2}, \ldots, m_{jm} \right] \times \left[w_1', w_2', \ldots, w_m' \right]^T$$

$$= \left[m_{j1}, m_{j2}, \ldots, m_{jm} \right] \times \left[\frac{1}{\sum_{h=1}^m h}, \frac{2}{\sum_{h=1}^m h}, \ldots, \frac{m}{\sum_{h=1}^m h} \right]^T$$

Calculate the forecasting value of every month from January 2021 to July 2021. Figure 2.10 represents the actual and forecasting values and shows the accuracy of the results with actual values.

We know that about the error if the value of error is smaller then, it talked about the accuracy of the model [8]. The root mean square error is calculated as follows,

$$RMSE = \sqrt{\frac{\sum_{i=1}^n \left(actual_i - forecasted_i \right)^2}{n}}$$

where n denotes the number of days needed for the forecast. Calculated RMSE value is 22.09.

2.19 CASE STUDY

To conduct an analysis of COVID-19 data with a time series model, forecast the data and check the fitness of analysis.

For the analysis of COVID-19 data, first, we consider two countries' data: India and Mexico. Consider that the data have some parameters, i.e., weekly confirmed cases, weekly confirmed deaths, cumulative cases, cumulative deaths, new cases, recovered data, vaccinated data, etc. For the analysis we consider the following parameters: weekly confirmed cases, weekly confirmed deaths, cumulative cases, and cumulative deaths. The World Health Organization (WHO) site was used for required data. The R-square and mean absolute percentage error (MAPE) was used to check the fitness of the model. The resulting 11 weeks of forecasted data for COVID-19 are shown in Figures 2.11 and 2.12.

$$MAPE = \frac{1}{n} \sum_{n}^{i=1} \left| \frac{x_i - \overline{x_i}}{x_i} \right| \times 100$$

If the high value of R-square and low value of error then the model is good in fitness. Values of R-square and error are shown in Table 2.2.

2.20 DISCUSSION

This case study discussed the COVID-19 data of India and Mexico. Also shown is the forecasting of the data for 11 weeks in Figures 2.11 and 2.12. Table 2.2 shows the numerical calculated R-square and error values, which demonstrate the fitness of the time series model in COVID-19. In the same way, we can apply the model to other countries, which is beneficial for studies and agencies in working to control viruses.

TABLE 2.2   Model Fitting Parameters for India and Mexico

	Model fitting parameters	India	Mexico
Last 7 days (new cases)	R- squared	0.981	0.761
	MAPE	9.166	15.936
Cumulative case	R- squared	1	1
	MAPE	0.62	0.56
Last 7 days (new death)	R- squared	0.961	0.592
	MAPE	11.39	26.75
Cumulative death	R-squared	0.999	1
	MAPE	0.542	0.627

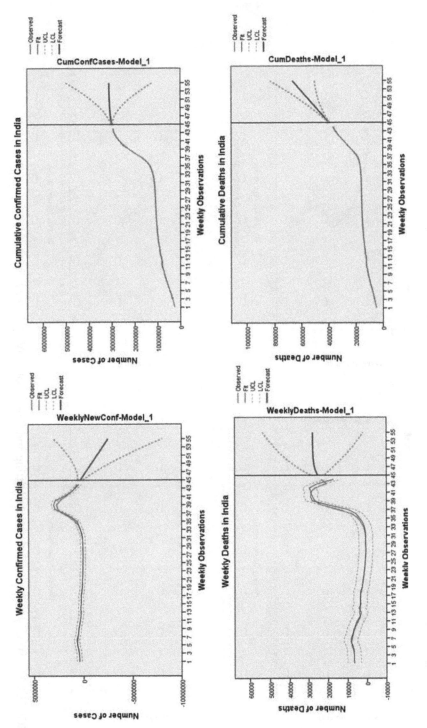

FIGURE 2.11 Weekly and cumulative model fitting and forecast of COVID-19 for India.

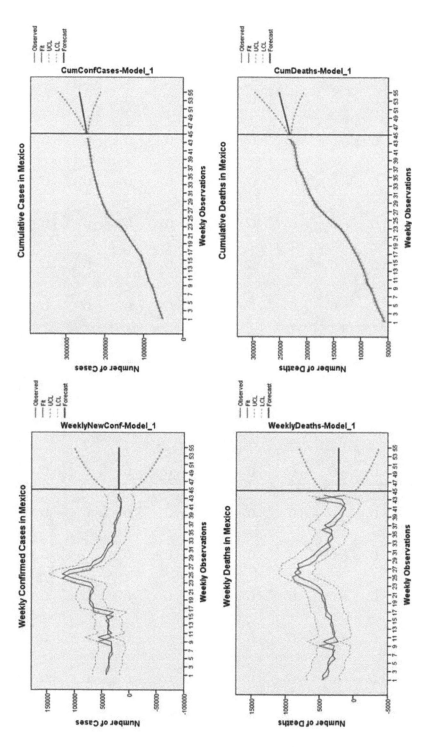

FIGURE 2.12 Weekly and cumulative model fitting and forecast of COVID-19 for Mexico.

REFERENCES

[1] Hyndman, R. J., & Athanasopoulos, G. (2018). *Forecasting: Principles and practice.* OTexts.

[2] Awasthi, A. K., & Garov, A. K. (2021). Quantity based weights forecasting for TAIEX. Vol. 6, pp. 405-409, *Journal of Physics: Conference Series, IOP Publishing.*

[3] Source of data; www.bseindia.com/Indices/IndexArchiveData.html.

[4] Yang, H., & Lee, E. K. (Eds.). (2016). *Healthcare analytics: From data to knowledge to healthcare improvement.* John Wiley & Sons.

[5] Awasthi, A. K., & Garov, A. K. (2021). *Agricultural modernization with forecasting stages and machine learning.* CRC Press.

[6] Awasthi, A. K., Garov, A. K., & Kumar, S. (2018). Integrating MATLAB and python. *Journal of Emerging Technologies and Innovative Research (JETIR),* 5(12), pp. 91–99.

[7] Chen, S. M., Chu, H. P., & Sheu, T. W. (2012). TAIEX forecasting using fuzzy time series and automatically generated weights of multiple factors. *IEEE Transactions on Systems, Man, and Cybernetics-Part A: Systems and Humans,* 42(6), pp. 1485–1495.

[8] Koul, S., Awasthi, A., & Garov, A. K. (2020). Experimental model approach for decision making in stock index. *Think India Journal,* 22(37), pp. 1272–1276.

CHAPTER 3

Intelligent Learning Analytics in the Healthcare Sector Using Machine Learning and IoT

Kundankumar Rameshwar Saraf
and P. Malathi

CONTENTS

DOI: 10.1201/9781003227595-3

3.1 INTRODUCTION

The intelligent health monitoring CPS presented by this chapter includes a bed, sensors, syslog server, one Splunk heavy forwarder, one Splunk indexer, and one Splunk search head. The syslog server and bed equipped with sensors are placed in a same single room. All other components are placed at remote location depending on the availability of physician and CPS admin. This system uses one bed equipped with various health monitoring sensors. Sensors generate the logs according to the patient's health condition. The bed is equipped with a temperature sensor, an oxygen level monitoring sensor, a blood pressure sensor, and a pulse rate sensor. Splunk detects abnormal health conditions based on the logs generated by these sensors. Splunk uses machine learning to perform this diagnosis. Splunk also uses artificial intelligence (AI) to anticipate the health issues of patients prior to their actual occurrence. This anticipation prevents major harm to the patient's body. Abnormal health conditions of patients are automatically sensed by Splunk and trigger the predefined alert. This alert will be received to physicians through phone calls, SMS, and email. After receiving the alert, the physician monitors the dashboard called "Smart Health Monitoring and Alerting System by Splunk", as shown in Figure 3.4. Physicians understand the abnormal health parameter using this dashboard. Depending on the abnormality, the physician writes a prescription for the appropriate medication, contacts the relative of the patient, and sends them the prescription.

To prevent major cyber threats to this system, Splunk monitors each CPS component. This system mainly deals with denial-of-service attacks and brute force attacks. On detection of these attacks Splunk sends an alert to the CPS admin via phone call, SMS, and email. CPS admins monitor the component for which the alert has been received. Depending on the cyber threat, the admin will take appropriate action and avoid further harm to the CPS.

3.2 PROBLEM DEFINITION

This section explains the need for intelligent health monitoring systems. Quick and accurate diagnosis of disease can save many patients'

lives. Skilled physician unavailability in villages, lack of health awareness, and high costs of diagnosis and medication lead to high mortality
rates.

Available IoT-based health monitoring systems can diagnose a health
issue only after its actual impact on a patient's health. These systems
may thus cause harm to the patient's health. Hence, intelligent health
diagnosis and predictive analysis of a patient's health is essential. The
system should detect the disease before its actual impact on a patient's
health.

This can be possible using the Splunk Industrial IoT platform. This software can predict the health issue before its actual occurrence. Early health
diagnosis using the automated intelligent health monitoring system can
reduce mortality rates. This system can detect the major health parameters
of a patient without actual intervention of physician. In emergencies and
unavoidable health issues, this system triggers an alert and sends it to the
physician.

3.3 LITERATURE REVIEW

Anish Hemmady (2014)[1] explained the significance of Splunk in healthcare and data security.

Healthcare institutions used Splunk Operational Intelligence features
in April 2015 [2].

Molina leadership team (Feb. 2014) monitored the patients' health by
using Splunk IT Service intelligence [3].

The Internet of Medical Things (IoMT) has predicted in 2017 that the
Splunk can be used for health monitoring and diagnosis [4].

Splunk use in big data analysis healthcare development frameworks was
explained by Venketesh P., et al. in October 2019 [5].

.conf19 (Dec. 2019) has presented the single dashboarding framework.
This framework monitors the health of athletes in real time [6].

According to Atef Kouki (March 2019) Splunk can be used to simplify
the diagnostic tasks of today's medical staffs [7].

Angelina P. K, (March 2021) has shown that Splunk can be used in
healthcare processes and to monitor all health parameters [8].

The white paper "Using Healthcare Machine Data for Operational
Intelligence" explained the use of Splunk to lessen billing errors
in hospitals, drug dispensing monitoring, and real-time fraud detection [9].

Medigate (https:medigate.pathfactory.com/medigate-and-splink) docu-
mented the use of Splunk to control medical and IoT devices [10].

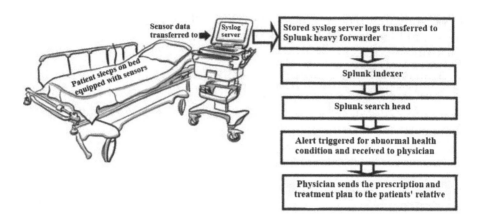

FIGURE 3.1   Intelligent health monitoring system.

3.4 BLOCK DIAGRAM OF SYSTEM

The intelligent health monitoring system is shown in Figure 3.1. This system is a Cyber Physical System (CPS) which has a bed equipped with various health-status monitoring sensors. The patient needs to sleep on this bed to diagnose themselves for any health criticality.

The sensors connected to this bed send the data to the syslog server. The syslog server collects the logs from sensors and transfer them to the Splunk heavy forwarder. The Splunk heavy forwarder parses the received logs and deletes unwanted logs. These parsed logs are then transferred to the Splunk indexer. The user has already created the indexes in the Splunk indexer. The Splunk indexer stores the logs in various indexes based on their type. The Splunk search head can monitor all the stored and indexed logs in the Splunk indexer. To monitor these logs, the user can create the dashboard and alert based on their requirements. All the devices shown in Figure 3.1 are connected via the internet.

3.5 HEALTH STATUS MONITORING SENSORS

In this CPS the bed is equipped with a weight sensor, temperature sensor, oxygen level monitoring sensor, and pulse rate measurement sensor. The following section explains the details of each sensor.

3.5.1 Temperature Sensor

This CPS uses a digital thermistor module to measure body temperature. This module contains a negative temperature coefficient (NTC) type thermistor and provides the output in analogue and digital format.

TABLE 3.1   Comparison of Temperature Sensors

Parameter	Thermistor	RTD	Thermocouple	Silicon
Accuracy	±0.1 °C	±0.01 °C	±0.5 °C	±0.15 °C
Ruggedness	Less affected by vibration and shock, most stable sensor	Can be damaged by vibration	Uses large gauge wire, hence rugged in nature	Its ruggedness is similar to dual inline IC package
Tempt. Range	-148 °F to 842 °F	-418 °F to 1652 °F	-454 °F to 3272 °F	-67 °F to 302 °F

Compared to other temperature sensors, the thermistor is as accurate, less susceptible to shocks or vibrations, and gives more stable readings in any adverse condition. Table 3.1 shows the comparison of all available temperature sensors.

3.5.2 Weight Sensor

At the bottom of the bed, the compression load cell is connected. To measure patients' weight, this load cell follows the principle of Wheatstone Bridge (strain gauge). Compression load cells are sturdy by nature. It is merely affected by unfavourable ecological conditions. The maximum load measuring capacity of this load cell is 1,000 kilograms (Kg).

3.5.3 Oxygen Level Monitoring Sensor

As shown in Figure 3.2, a pulse oximeter is used by this system to measure the oxygen level, pulse rate, and PI index of patient. Based on transmissive technology, this sensor measures the oxygen level in human body. Patients place their middle finger in the pulse oximeter. The SpO2 level (oxygen saturation level) of blood is measured by the sensor and shown in the form of a percentage. For normal human beings, the SpO2 level is always above 94 percent, below which a patient requires emergency treatment.

The ratio of pulsatile blood flow to static blood flow is called PI. It is one of the important parameters to measure and show the patients' pulse strength. PI value varies from 0.02 percent to 20 percent. Less than 4 percent PI is usually treated as an alarming situation for the patient's health and requires urgent treatment and medication. The pulse oximeter shows the PI value in percentages.

The per-minute beat rate of normal human beings varies between 70 and 100. The pulse oximeter also shows the value of heartbeat rate per minute in the form of beats of per minute (BPM). BPM values less than 70 and more than 100 indicate an unfavourable health condition.

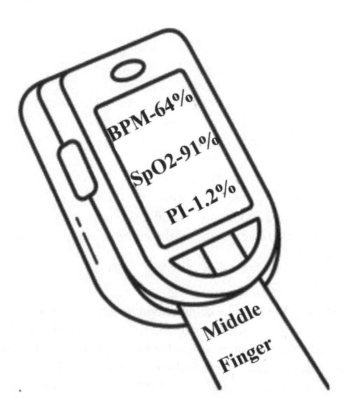

FIGURE 3.2   Pulse oximeter.

As shown in Figure 3.2, the pulse oximeter shows various health parameters of the patient. According to the readings in the figure, it can be concluded that patient needs emergency treatment.

3.6 SPLUNK DASHBOARD TO MONITOR PATIENT HEALTH STATUS

To diagnose the health of a patient, this research has created a dashboard with six panels. The first panel of this dashboard shows the general details of a patient (as shown in Figure 3.3). These general details include name of patient, gender, age, contact details, and three options to enter their medical history. These details are directly saved to the syslog server, i.e., a computer placed near the bed of patient. General details can be filled in by the patient or their relatives. Sensors connected to the bed measure the physical health of the patient and store every parameter in the syslog server. This server sends general and physical patient details to the heavy forwarder. The heavy forwarder performs data filtering and forwards the filtered data to the indexer. CPS admin can search the indexed data by writing various specified Splunk queries on the Splunk search head. CPS admin can create a dashboard to visualize the indexed data by syslog server.

The authors have created in this chapter a dashboard named "Smart Health Monitoring and Alerting System by Splunk". This dashboard has six panels, as shown in Figure 3.3. The first panel shows the general details of patient, such as name, gender, age, contact details, and medical history. The second through sixth panels of this dashboard show the physical health status of the patient measured by sensors. This status includes the following parameters: weight, body temperature, blood oxygen level, pulse rate, and perfusion index.

As shown in Figure 3.3, the patient named Harshal Patel, aged 34 years, has a medical history of asthma and high blood pressure, as shown in the dashboard on panel 1. Dashboard panel 2 shows that the patient's weight is 72 kilograms. This weight is measured by load sensor. The patient's body temperature is 98 °F, which is normal and shown in the third dashboard panel. This body temperature is measured by digital thermistor temperature sensor module. The SpO2 level of the patient is 97 percent, as shown in the fourth dashboard panel. The heart rate is 79 BPM, as shown in the fifth dashboard panel. The perfusion index (PI) is 5.8 percent, as shown in the sixth dashboard panel. The PI level, SpO2 level, and heart beats are measured by pulse oximeter.

As shown in dashboard panels 2 to 6 of Figure 3.3, all the present health conditions of patient Harshal Patel are normal. Hence, no alert will be sent to the physician. The Splunk suggests some usual medicines to solve

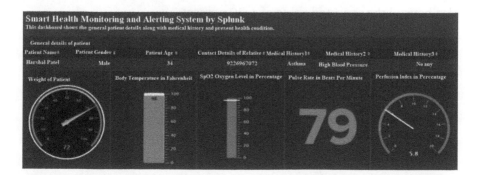

FIGURE 3.3   Smart Health Monitoring Splunk Dashboard for Harshal Patel.

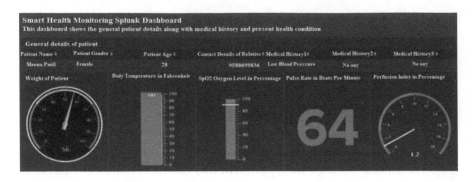

FIGURE 3.4   Smart Health Monitoring Splunk Dashboard for Meena Patil.

normal health issues faced by Harshal, such as headache or body pain. These usual medicines are already stored by physicians to cope with the normal health issues faced by patients.

Figure 3.4 shows the dashboard with the health status of one female patient. This dashboard also contains the similar six panels. The first panel of this dashboard shows the general details of the patient. By this dashboard one can know that patient's name is Meena Patil, who is 28 years of age and who has a medical history of low blood pressure. The other panel lets one know that Meena weighs 56 kilograms and has a body temperature of 101 °F. Her oxygen level (SpO2) is 91 percent with a heart rate of 64 BPM. Meena's PI is 1.2 percent. As per the data received by the third through sixth panels of the dashboard, it can be concluded that Meena needs urgent, lifesaving medical treatment. In this case, the body temperature is beyond the predefined temperature range (above 98.5 °F). The SpO2 level is also beyond the predefined range (below 94 percent). The heart rate has also crossed its threshold limit of 70 to 100 BPM. Perfusion index is also beyond the expected value of 4–10 percent.

3.7 SPLUNK ALERT TO NOTIFY PHYSICIAN OF EMERGENCY HEALTH ISSUES

Figures 3.5 to 3.8 show critical health conditions of patients which trigger the alert to notify the physician of an emergency situation. This alert sends notification of emergency to physicians by phone call, SMS, and email. On receipt of this alert, physicians observe the health monitoring dashboard. Based on the parameter variation of this dashboard, physicians suggest prescribed medication and treatment plans to the patient's relatives.

This treatment can immediately control the critical health condition of the patient.

Splunk alerts have three different parts, as follows: settings, trigger conditions, and trigger actions. Settings and trigger conditions are shown in Figure 3.5. Settings includes title, description, permissions, alert type, and expire time of alert. The detailed explanation of each of these options is shown below:

- **Permission:** The alert can only be used by alert creator if the 'private' permission of alert is selected. Alert can be used by all users and all apps if the 'shared in app' permission of the alert is selected.

- **Alert type:** This indicates the search timing of the alert. Scheduled alert type includes various options such as 'run every hour', 'run every day', 'run every week', 'run every month', and 'run on cron schedule'. Users can select any one of the options to run the alert search with the scheduled option. Real-time option alert type indicates that the alert search will continuously run after its creation.

- **Expires:** This option shows the lifespan of every triggered alert. It indicates the period for which the result of the triggered alert can be observed. As shown in Figure 3.5, the alert will expire after 24 hours. This means that the triggered alert can be observed for up to 24 hours after its occurrence.

Using the trigger conditions option of Splunk alerts, the user can prioritize the events that occur during the alert triggering. Description of various options of the trigger conditions section of the Splunk alert is shown in the following:

- **Trigger alert when:** This option, shown in Figure 3.5, includes number of results, number of hosts, number of sources, and custom fields. On selecting the number of results options, users can decide the alert triggering after 'x' number of similar events generated by the alert search. The number of hosts option indicates that the alert will be triggered only after a certain number of hosts generate the event mentioned in alert search. The number of sources option indicates that the alert will be triggered only when a certain number of sources generate the events mentioned in an alert search. The custom option

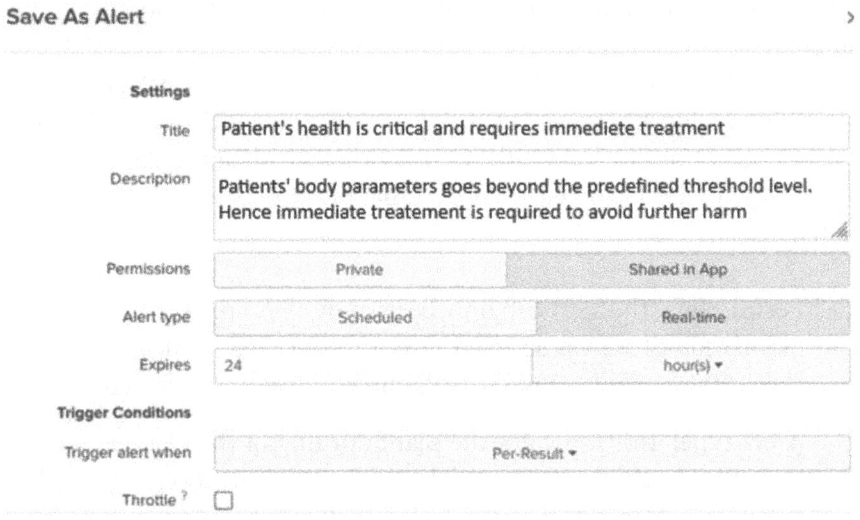

Save As Alert >

Settings

Title	Patient's health is critical and requires immediete treatment
Description	Patients' body parameters goes beyond the predefined threshold level. Hence immediate treatement is required to avoid further harm
Permissions	Private	Shared in App
Alert type	Scheduled	Real-time
Expires	24	hour(s) ▼

Trigger Conditions

Trigger alert when	Per-Result ▼
Throttle ?	☐

FIGURE 3.5   Settings and trigger conditions parts of alert.

of trigger alert indicates that the user can select any condition; for example, when the search count is more than ten, the alert should be triggered. The 'trigger once' option indicates that the alert will be triggered only once the threshold condition is reached. The 'for each result' option indicates that the alert will trigger after every result of the expected alert condition is reached. For example, if the user has selected the 'for each result' option of an alert trigger, the alert will trigger whenever the mentioned search condition is met by the search result. So, for 1,000 search results, 1,000 different emails will be sent to the user. The trigger once option sends a single email which includes 1,000 results in it.

- **Throttle:** By selecting this option, the user can suppress the subsequent alert for a certain duration of time after the occurrence of each alert.

Figure 3.5 below shows that this alert is created to trigger after the detection of every emergency (per result basis). Hence, the physician is notified for every abnormal health condition of a patient. The health status of all patients is continuously monitored by this alert in a real-time manner. This alert will expire after 24 hours of its triggering. It is not suppressed for any specified duration; hence the throttle option is unchecked.

Trigger actions shown in Figure 3.6 include the various actions taken upon the triggering of each alert as given in the following:

- **Add to triggered alert action:** This indicates that the alert will be reflected in the triggered alert section of Splunk.

- **Log event:** The user can construct the custom log events to search and index the metadata. This option indicates that on occurrence of every triggered alert, the log events will be sent to the Splunk receiver endpoint or deployment.

- **Output results to lookup:** This shows that the triggered alert results will be saved to the CSV lookup file on occurrence of alert.

- **Output results to telemetry endpoint:** This option sends usage metrics back to Splunk.

- **Run a script:** This option will run the predefined python script on occurrence of alert.

- **Send email:** This option sends the email notification to the selected user on occurrence of alert.

- **Send to Splunk Mobile:** This option sends the notification to the users of Splunk on their smartphones.

- **Webhook:** This option sends the HTTP POST request to specified URLs on occurrence of alert.

In Figure 3.6, the user has selected the send email option to send email to the selected recipient. Splunk sends the email to the physician and notifies the physician of the emergency. The physician will observe all dashboard panels related to the received the alert. By observing all these panels, the physician comes to know the criticality of the patient's health.

Figure 3.7 shows the webhook option of the Splunk alert. This option sends the HTTP POST request to www.zenduty.com. This website sends email, SMS, and phone call notifications to the physician. The user needs to insert the physician contact details on this website to create the token. This token is inserted in the URL option of the webhook alert action.

Figure 3.8 shows that the 'add to triggered alerts' action is selected by the user. This trigger action saves the alert notification in the triggered alert section of the Splunk user interface. The physician and Splunk user can count the total number of received alerts using this option.

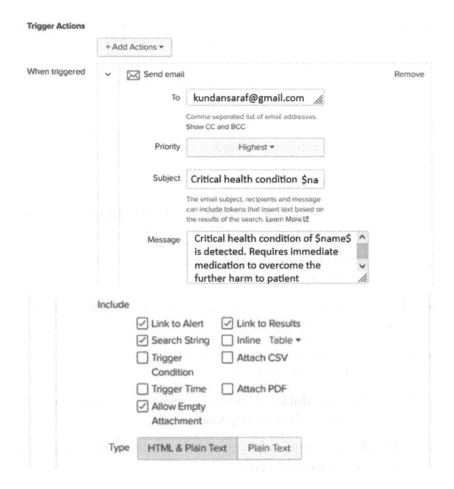

FIGURE 3.6   Trigger action of alert configured to send email to physician.

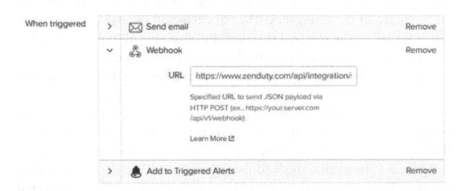

FIGURE 3.7   Webhook option configuration for phone call and SMS alerting.

FIGURE 3.8   Triggered alert configuration of Splunk alert.

3.8 INTEGRATION OF HEALTH MONITORING SYSTEM WITH SPLUNK FOR INDUSTRIAL IOT

This CPS can be integrated with Splunk for industrial IoT. This integration helps perform predictive analysis of health dashboard and security attacks. Splunk for industrial IoT monitors the health status of patients; any unexpected behaviour is recognised by this app and triggers a notified event 30 minutes before its actual occurrence. Physicians can see the notified event, refer to the dashboard, and suggest medication to avoid any future harm to the patient's health. In a similar way, the Splunk for industrial IoT app detects future cyber threats 30 minutes before their actual occurrence. On detection of any cyber threat by predictive analysis method, this app sends the notable events to the CPS admin. This admin takes necessary action after viewing the notable events and avoids future cyber threats. Cyberattacks considered by this research are outlined in the following sections.

3.8.1 Denial of Service of Attack

Denial of Service (DoS) attacks occur when the traffic on any application, website, or network increases above its capacity. This attack creates congestion in the communication. Splunk for industrial IoT monitors the usual network traffic of the CPS. Whenever the traffic increases above the normal per-day limit, it triggers a notification to the CPS admin. On reception of this alert, the CPS admin will take required actions to avoid the further harm to the CPS.

3.8.2 Brute Force Attack

In a brute force attack, the attacker tries to gather the login credentials of the CPS component. For recognising these credentials, the attacker enters multiple combinations of usernames and passwords. Splunk for industrial IoT will monitor the number of failed login attempts. Once these attempts go above the predetermined threshold limit, it triggers the event and informs the CPS admin. After receiving this notification, the CPS admin can verify the authenticity of the login. If the failure attempt is performed by a legitimate user, the admin will close the alert. Otherwise, the admin will block the IP address from where the alert failure login attempt has originated.

3.9 ADVANTAGES, DISADVANTAGES, AND APPLICATIONS OF THE SYSTEM

3.9.1 Advantages

a) This system can monitor the health condition of patients without intervention of physicians.

b) This system helps physicians monitor the health of multiple patients simultaneously.

c) The patient situated in a remote location can get timely treatment by a good physician.

d) By providing on-time accurate diagnosis, this system highly reduces the mortality rate of patients.

e) This system can secure the entire CPS against cyberattacks.

f) One-time investment costs are required to implement this system, which may diagnose a huge number of patients. Hence diagnosis by this system is much cheaper for the patient.

3.9.2 Disadvantages

a) This system completely depends on the internet and electricity. Hence, the failure of internet connections or electricity can interrupt the whole functionality of the system.

b) This system's diagnosis depends on sensors. Malfunctions of any sensor leads to inaccurate readings by the system.

c) To check the accuracy of this system, monthly calibration of sensors and other system parts is essential.

3.9.3 Applications

a) This system mainly is used for patients situated in remote locations.

b) This system can be used during pandemics to treat a high number of patients.

c) This CPS can be useful to diagnose patients affected by contagious diseases like COVID-19. Normal symptoms of COVID-19 patients include headache, body pain, low oxygen level, and fever. In this system, the user stores contact details including email ID, and phone number. Physicians set up the threshold of all health parameters. The sensors connected to the bed measure all health parameters of the patient. Splunk triggers alerts and notifies the physician if any parameter goes beyond the threshold level. Physicians can send prescriptions to patients' relatives. Splunk also detects the DoS and brute force attacks on the various CPS components. In case of any unusual cyber threat, this CPS triggers alerts and notifies the CPS admin. This admin can overcome the cyber threat on the CPS component.

3.10 CONCLUSION

This chapter explains the intelligent health monitoring system. The dashboard and alert configured by this system are briefly explained. Integration of this system with Splunk industrial IoT performs the predictive analysis and sends the alert or notified event to the physician before the occurrence of any future health issue. This system can also monitor various cyber threats to the CPS. Before occurrence of any harm to the system by these cyber threats, this system notifies the CPS admin. This system is extremely useful for patients located in remote locations and who have a scarcity of good medical facilities. During pandemics like COVID-19, the number of patients is high. Hence available physicians are insufficient to provide good treatment to all the patients within a stipulated time. This system is useful in such a situation. By using this system, physicians can monitor multiple patients and suggest essential medications to patients with serious health issue. Physicians can diagnose large numbers of patients and concentrate on the health issues of only emergency patients. This system can highly reduce the mortality rate and increases health awareness. This system is also relatively cheap and enables any patient to check their health condition.

ABBREVIATIONS

AI:	Artificial Intelligence
BPM:	Beats Per Minute
COVID-19:	Corona Virus Disease 2019
CPS:	Cyber Physical System
DC:	Degree Celsius
DF:	Degree Fahrenheit
E&TC:	Electronics and Telecommunication Engineering
IoMT:	Internet of Medical Things
IoT:	Internet of Things
IP:	Internet Protocol
IT:	Information Technology
KG:	Kilo Gram
NTC:	Negative Temperature Coefficient
PI:	Perfusion Index
RTD:	Resistance Temperature Detector
SMS:	Short Message Service
OSL:	Oxygen Saturation Level

REFERENCES

[1] Anish Hemmady, "Significance of Big Data on Healthcare and Data Security", *International Journal of Scientific & Engineering Research*, Volume 5, Issue 12, December 2014.

[2] Press release by Splunk on "Splunk Builds Strong Traction in Healthcare", 2015 www.splunk.com/en_us/newsroom/press-releases/2015/splunk-builds-strong-traction-in-healthcare.html

[3] Press release by Splunk on "Splunk IT Service Intelligence Helps Molina Healthcare Deliver Innovation to Patients", San Francisco, 21st February, 2017 www.splunk.com/en_us/newsroom/press-releases/2017/splunk-it-service-intelligence-helps-molina-healthcare-deliver-innovation-to-patients.html

[4] Data insider, "What Is the Internet of Medical Things (IoMT)?", 2017 www.splunk.com/en_us/data-insider/what-is-the-internet-of-medical-things-iomt.html

[5] Venketesh P., and Ramkumar T., "Implications of Big Data Analytics in Developing Healthcare Frameworks: A Review", *Journal of King Saud University: Computer and Information Sciences*, Volume 31, Issue 4, Pages 415–425, October 2019.

[6] Splunk, "Reintroducing Splunk Dashboards", 2019 www.splunk.com/en_us/blog/platform/reintroducing-splunk-dashboards.html
[7] Splunk, "Splunk at the Service of Medical Staff", 2021 www.splunk.com/en_us/blog/platform/splunk-at-the-service-of-medical-staff.html
[8] Angelina P. K, Geoff H, David H, and Owen J, "Process Mining on the Extended Event Log to Analyse the System Usage During Healthcare Processes (Case Study: The GP Tab Usage during Chemotherapy Treatments)", *International Conference on Process Mining, ICPM 2020: Process Mining Workshops*, pp. 330–342, 31st March 2021.
[9] White paper, "Using Healthcare Machine Data for Operational Intelligence", 2012 https://davidhoglund.typepad.com/files/splunk_for_healthcare.pdf
[10] Medigate S, "Clinical SOC Solution Delivering Complete Visibility and Control Over Medical and IoT Devices", 2021 https://medigate.pathfactory.com/medigate-and-splunk

CHAPTER 4

Influence of AI and Machine Learning to Empower the Healthcare Sector

Sumit Koul, Bharti Koul, and Bhawna Bakshi

CONTENTS

DOI: 10.1201/9781003227595-4

4.1 INTRODUCTION

Industry 4.0 is mainly destined for the healthcare sector, which is known as healthcare 4.0. The improvement in the healthcare sector is required to be done at various levels by the use of IoT and automated AI techniques. The existing healthcare system is lacking in different subsectors of healthcare to provide appropriate treatment to the patients. The key factor for enhancing the efficiency of healthcare sector is the incorporation of industry 4.0. Newly emerging technologies have gained popularity as a way to enhance the healthcare sector. In today's world, a huge amount of information is collected from patients when they register or enter hospitals, as well as when these patients are discharged from the hospitals. Such information is crucial for those who are working and performing research in this field. Technologies like big data, Internet of Things (IoT), blockchain, and machine learning (ML) algorithms have a key role to play for the improvement of the healthcare system. Because of the huge amount of prevailing data regarding the patient's diagnosis and medication, the pressing issues are firstly how to upload all the information and secondly how to safeguard the patient's data. In this regard, big data analytics play an important role in the safety and security of information related to the healthcare system. Firstly, collecting information manually as was done conventionally is an old technique. Today information is collected via IoT-based techniques which collect information from the patients digitally. Blockchain is garnering attention by both the researcher and scientist. ML algorithm helps medical practitioners to focus on those issues which have not been tackled previously.

Another technology that is gaining attention among the researchers and scientists is blockchain technology. To safeguard the personal information of patients is of top priority. Blockchain technology can provide better solutions for it.

Aceto et al., 2020 briefly explained the concept of big data, IoT, and fog and cloud computing to enhance the healthcare sector which is pushing towards industry 4.0. They have shown that how these technologies have empowered old methods used in healthcare sector, which is being highlighted in industry 4.0. Ajmera and Jain, 2019 have focused on the 15 barriers which influence healthcare 4.0, and they make some suggestions for ways to overcome the problems therein. Introducing the blockchain concept in the healthcare sector adds efficiency and makes it more automated as well as increases the managerial skills of healthcare sector.

This technology is transparent so that no security leakage can be made about patient information, which is collected electronically (Farouk et al., 2020). Maintaining electronic health records is very important today for diagnostics and analysis of patient ailments. It is also important to explain different types of existing methods which can control the security threats as well as challenges in handling patient's records, etc. Also, in comparison to old methods, a new method is proposed that manages health record electronically so as to overcome the drawbacks of old methods. This proposed method helps medical practitioners get information about patients from any corner of the world (Hathaliya et al., 2019). Tortorella et al., 2020 study the various issues, research gaps in existing theory, and trends to overcome the drawbacks in current healthcare 4.0. Hathaliya et al. have proposed a new method which overcomes the various drawbacks of healthcare 4.0 to provide the solution to government and private hospitals. YIN et al., 2016 have emphasised the role of IoT with respect to the healthcare sector and what IoT can do so to enhance the capability of this sector. Saheb and Izadi, 2019 have described the three phases of IoT with big data which can handle the storage of records, security with regard to patent information, and data processing. IoT with big data technique can provide better results in enhancing the healthcare sector. Mehta et al., 2019 has considered a literature review to come up with conclusions on future research on artificial intelligence (AI) as well as big data to transform the healthcare sector in the current era. Ojokoh et al., 2020 have explained the three important concepts of AI, big data, and analytics, which laid the foundation of new healthcare. These three explore the many hidden sectors in healthcare which contain challenges and also can handle the serious issues. Waschkau et al., 2019 have discussed the big data technique and provided a better solution for controlling multimorbid patients' data. Waschkau et al. have also shown that less literature work has been done as of yet by researchers or academicians. An extensive use of data mining techniques in the healthcare sector has rapidly increased the current situation. Islam et al. (2018) have explained the role and importance of data mining in healthcare by an exhaustive review of literature. A technique is developed to monitor the patients remotely. This technique resolves the issue of a patient's ailment which is cost efficient, reliable, securable, and practical within time constraints (Kakria et al., 2015). Tan and Halim, 2019 have proposed IoT-based monitoring devices which collect patient's vital information such as body temperature, blood pressure, and the pulse rate. The main function of the device is to receive the information for patient and send remotely to the medical practitioners. By observing the data of patient remotely, they can suggest medications to the patient. By this, patient can be treated from home, and it can save time as well as money. Kadhim et al., 2020 focus on applications of IoT in the healthcare system which upgrade the conventional procedures that have been followed in the medical field for decades. Malasinghe et al. (2019) have discussed the monitoring of the patients remotely using sensor. Malasinghe et al. has shown significance results in remote monitoring of patients as compared with the conventional practice in the medical field. This would be an enhancement in healthcare sector which is known as industry 4.0.

This chapter is organized as follows: Section 4.2 describes IoT and healthcare and explains the benefits of IoT as well as wearable technologies used in the healthcare sector. Section 4.3 discusses big data and its role on in healthcare. Machine learning and its benefits for the healthcare sector is considered in Section 4.4. In Section 4.5, three different case studies are considered that explain the role of IoT to empower the healthcare sector today and in the future. Section 4.6 concludes with this chapter offering a future scope.

4.2 IOT AND HEALTHCARE

IoT-based technologies have played a significant role in the healthcare sector. With the influence of IoT, the patient is attended by medical practitioners remotely if the patient does not have a serious medical problem. In such cases the sharing of information is performed from both sides. On the patient's end the symptoms of disease are shared, and from the doctor's end the diagnosis and treatment are shared. So by using IoT-based technology patient problems can be resolved remotely. Constructive decisions can be made on the bases of the information collected remotely by medical practitioners form the patient. Such interactions by the help of IoT will transform the healthcare sector. Here all the participants are benefitted in one or another way. Like taking the case of a patient whose expenditure, the hospital bills along with the doctor's consultation, fee is to be borne by the insurance company, in such section it is patient oriented IoT and the bills are paid by the insurance company Medical practitioner oriented IoT, Hospital oriented IoT as well as Health insurance oriented IoT. Figure 4.1 represents the different section who is benefit form IoT gadgets.

IoT has its own architecture in the healthcare sector. IoT is divided into four stages, as described in Table 4.1.

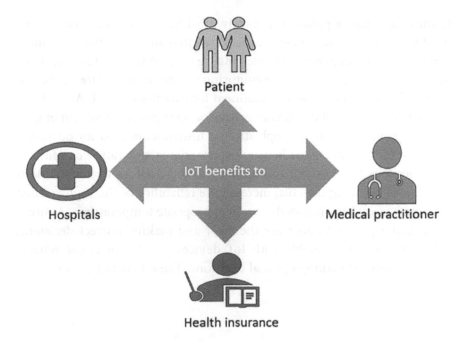

FIGURE 4.1   Applications of IoT.

TABLE 4.1   Stages of IoT

Stages	Description
Stage 1	Collecting patient's information with the help of sensor-equipped devices to record temperature, heartbeat, pulse rates, as well as O_2 levels. Laser-guided cameras to capture body aliments for further analysis.
Stage 2	Since the gathered information which is collected form sensor-equipped devices is in raw form it needs to be converted into group form. For this, an aggregation method is applicable at this stage.
Stage 3	When the aggregation process is done. A huge amount of information is sent to the data warehouses or cloud-based system to keep records of information for future analysis.
Stage 4	The data is kept under surveillance in data warehouses. Such data cannot be utilized by everyone; only authorized persons can use it.

IoT provides various cost benefits by minimizing the transportation expenses of the patient. Doctors can monitor the patient remotely which can benefit the patient and give significant results. Enhanced treatments mean that doctors can provide better solutions to patients while taking optimal decisions as well as maintain reliable transparency in prescribing the medications. IoT-based devices provide the rapid detection of disease in the cases where a patient is critically ill with some aliment and doctors want to see the seriousness of the disease so that an appropriate treatment can be started. Various IoT devices can be used to resolve these types of issues. An ongoing method of monitoring various issues related to health has given rise to the proactive treatment for patients in need. A challenging task for any hospital, whether private or semi-government run or government run, is to manage sophisticated instruments used for operation as well as for preserving medicines at appropriate temperature conditions. This increases the cost of running the hospitals. IoT can offer solutions in terms of reducing the cost, increase the reliability of the sophisticated instruments, and storing medicines at appropriate temperatures. Another important aspect is to decrease the error and making correct decisions. Again, this can be possible with IoT devices which can create reliable environments for making optimal decisions. These benefits are shown in Figure 4.2.

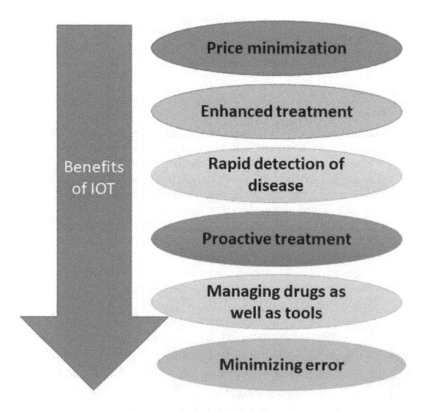

FIGURE 4.2   Benefits of IoT in the healthcare sector.

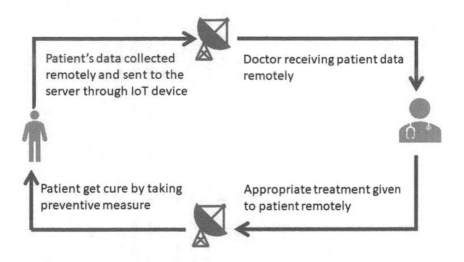

FIGURE 4.3   Remote patient treatment.

4.2.1 IoT-based Monitoring Devices in the Healthcare Sector

It is projected that by 2026, $176 billion profit will be made by IoT companies in the healthcare sector. A rapid growth in the healthcare sector will enhance various subsectors as well as medical devices which increases the rate of growth of IoT companies. The term that is commonly used is the Internet of Medical Things (IoMT). Some of the common and reliable IoT devices which can explain the concept of IoMT in the healthcare sector are described in Table 4.2. How monitoring device work is shown in Figure 4.3.

4.3 BIG DATA AND HEALTHCARE

A huge amount of data is coming into the medical sector. The problem arises from the security of the data and how to keep it safe from unknown persons. Also, storing such large amounts of information is another big challenge. Conventionally, data is collected manually and to store it is a tedious task. With the new techniques data is collected and stored digitally. This is known as an electronic health record (EHR). An EHR contains different types of metrics. These metrics are patients' particulars, laboratory reports, medical practitioner reports, observed function of apparatus used in the hospital for patients, insurance policy, etc. Interpreting with these metrics is not possible with usual defined software. Also, even if the data is collected digitally, but the main issue is to store the data safely. This issue is resolved with the

TABLE 4.2 IoT Devices and Their Major Issues

S. No.	IoT gadget	Application	Major Issue
1	Observing patients remotely	The easiest and most common application in healthcare is to observe patients remotely through IoT apparatus. IoT gadgets digitally collect information of patients. This information is then stored digitally in the data warehouses through special software which has a peer-to-peer network connection. For instance, the data on the bases of heart rates, systolic or diastolic pressure, temperature, blood sugar, etc. is collected from patients every hour by IoT gadgets remotely and sent to the medical practitioner through sophisticated apps or software to enable continuous remote observation. This also allows the practitioner to suggest required treatments to the patient quickly, which saves the patient time as well as money.	When the data is collected remotely by the IoT apparatus then the major issue is to keep such information safe from unknown persons.
2	Observing sucrose levels remotely	A large population of the world faces the very serious chronic disease of diabetes. Those patients who are suffering from diabetes have to adjust their daily routine work whatever they were doing in past, so that blood glucose can be normalised. Such changes include taking hour-long walks, proper medication, an hourly or daily basis routine check of sucrose level in the body, taking meal in piecewise way, etc. For the purpose of checking the sucrose level in the patient body, he or she has to go to a blood collection centre and give blood. After a few hours he or she can collect the report. Then the patient shows this report to the medical practitioner who gives the proper medication to the patient and suggests some other appropriate measures. This is the old treatment method from diagnosis to proper medication. Now IoT gadgets can offer a solution by reducing the time gap as well as other hardships on the part of the patient. Now, the patient can collect his or her sucrose levels remotely by highly modified IoT gadgets. These gadgets store information and connect with an app. Such information can be easily seen by medical practitioners remotely. This makes it easier for both the patient as well as medical practitioner, saving time and money.	Various issues can be solved easily by IoT gadgets. Like digital records of a patient. IoT gadgets not only check the level of sucrose but also can alarm the patient time to time about the level of sucrose. In addition to this, IoT companies are making smart gadgets for patients which have decent features in it, such as causing less consumption of electricity, giving quick responses as well as portability, and providing reliable results.

| 3 | Observing depression as well as mood of patient | Analysis the mood as well as depression of physically ill patients. A medical practitioner who is treating patients for aliments like mood as well as depression has to observe his or her reaction for some interval of time. For this purpose, an IoT gadget gives solution to the generic problem faced by the medical practitioner to observe the behaviour of patient. Medical practitioner remotely can judge the behaviour of the patient and suggest more reliable treatment to the patient. This will benefit the patient as well as the medical practitioner. IoT gadgets can overcome the drawbacks of existing systems in medical sectors. | Issues like mood analysis as well as guessing depression through IoT gadgets can eliminate the gap between the patient and the medical practitioner. Smart gadgets are developed by IoT companies which are less in cost, give reliable results, maintain the accuracy, and also provide better results. |
| 4 | Operation through robots | Today, instead of human involvement in operating rooms, various hospitals use robots to do complex surgeries. Instructions are given to the robots by doctors remotely. Within a small span of time, the surgery is over and patient can recover as well as get discharged from the hospital. This is how IoT can do better in the healthcare sector. | IoT companies are developing smart robots which can be used for complex surgeries. Implementation of equipment in hospital as well as training the medical staff is important. It takes time and cost for enhancing the system. |

help of big data analytics (BDA). The potential of BDA is quite impressive to handle many issues in the healthcare sector. An exhaustive analysis of a huge amount of the patient's data is to be carried out for gathering more input as well as to gain more understanding of it. This will help the medical practitioner get an information regarding patients with different ailments in advance. Medical practitioners can solve complex problems and come out with an optimal decision which will be favourable to the patient. The role of BDA increases the quality, reliability, and accuracy of results, decreases risk, and controls the access of huge amounts of data, organizing it for the purpose of analysis as well as its interpretation. Using these metrics BDA studies not only the past but also lays down the roadmap for the future, which gives an advantage to those who are in the field of medical sciences.

Healthcare analytics (HA) is divided into four categories: descriptive, diagnostic, predictive and decision analytics. Cleaning as well as summarisation of EHR comes under the descriptive category. To comprehend the past, the descriptive category gives a valuable outcome. The main goal is to see what happened earlier. Hindsight gives us an ideal of the past but cannot make complete decisions for healthcare providers. Analysis of raw data comes under diagnostic analytics. For instance, a medical practitioner wants to know the severity of the disease in patients. But some diagnostic analytics give the wrong outcome due to inefficiency in raw information that was previously collected. Predictive analytics work similarly to diagnostic analytics with a slight change. Suppose the medical practitioner knows the seriousness of a patient's aliment. If the medical practitioner wants to analyse the future diagnosis as well as treatment provided to the patient, the predictive analysis can provide better solution to it. Prescriptive analysis took correct decision about the patient. Figure 4.4 represents big data as well as healthcare analytics to provide evidence-based solutions.

HA provides enhanced healthcare systems which benefit all players such as medical fraternity; civil society; the one-to-one relation between medical practitioner and patient, and companies who are making IoT gadgets used in robotic surgeries, monitoring patient remotely as well as pharmaceutical. A distinguish feature of HA has many layers such as minimizing risk, increasing the opportunity for advanced research programmes, advanced outcomes, increased levels of competition and reliability of products. These all need back-to-back cutting-edge software which converts ungroup information to group information and non-actionable to

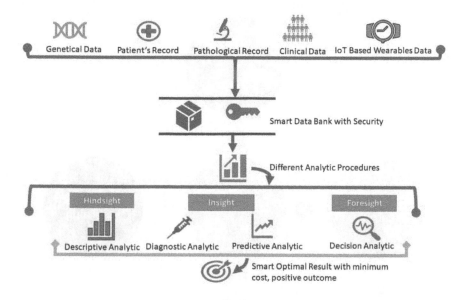

FIGURE 4.4   Smart architecture big data analytics to healthcare analytics.

actionable which connect real-world problems to evidence-based out-comes. For doing these, following steps should be considered: gathering correct information, transmuting the gathered information, visualising as well as treating analytical to achieve goals. In addition to this, we have to use cutting-edge technology for the smooth functioning of software which can provide evidence-based outcomes.

4.4 MACHINE LEARNING AND HEALTHCARE

In the present scenario digitization plays an important role in almost all of the world, such as in corporations, businesses, trades, industries etc. It is becoming essential for all of us to share, capture, and deliver informa-tion. It provides transparency in ongoing system. One of the industries is healthcare which need surplus power to handle the huge amount of data so that it can manage in right way. Techniques like BDA, HA, and machine learning techniques (MLT) can handle issues related to the healthcare sec-tor, and whatever the challenges are these techniques provide solutions for optimal decision-making. Different issues related to healthcare sector such as minimizing the cost, familiarising with cutting-edge technologies etc. For the smooth function of highly sophisticated software or equipment are not addressed properly in ongoing system. MLT give varieties of techniques

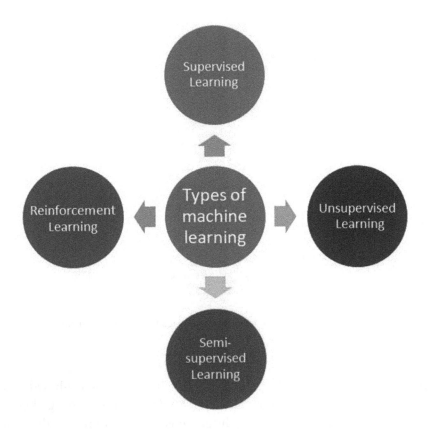

FIGURE 4.5   Different types of ML.

by which one can get innovative ideals to enhance healthcare sector. ML is providing solutions in healthcare which cannot be solved with ongoing procedures. These advantages make ML a sound, authenticated, securable system which can enhance healthcare. The ongoing system of healthcare contains many loopholes. ML has an ability to overcome these loopholes and come up with improved, reliable systems or solutions. Types of ML are described briefly (Koul, 2021) in Figure 4.5. These types of ML contain algorithms which provide reliable results with minimized cost as well as evidence-based study.

4.4.1 History of ML

The history of ML is shown in Figure 4.6. Table 4.3 explains the evolution of ML. Table 4.4 provides the brief history of ML in the healthcare sector. A comparison has been made in Tables 4.3 and 4.4 which shows a great achievement in healthcare since 2000.

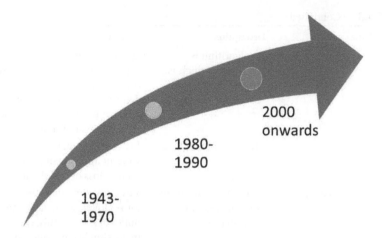

FIGURE 4.6   Historical journey of Machine learning.

TABLE 4.3   History of ML Innovations

Year	Name	Description	Application
1943	Warren McCulloch and Walter Pitts	Developed neural network (NN) while working on electrical circuits	
1952	Alan Turing	World famous Turing test	
1952	Arthur Samuel	Developed programming on computer for first time.	Initiating the computer programming.
1958	Frank Rosenblatt	Develop artificial neural network (ANN) for first time. It is also recognized as perception.	Main achievement is to recognized the shape as well as pattern.
1959	Bernard Widrow and Marcian Hoff	A modified version of neural network is introduced.	Two useful applications are developed. One is ADELINE and another one is MADELINE. Former one is used for finding the binary pattern. The main achievement is to predict the future bit sequence which has not been found before it. Later one is MADELINE. This model work for echo problem occurs in telephonic lines. This development is useful for the whole world.

(Continued)

TABLE 4.3 (Continued)

Year	Name	Description	Application
1967		An algorithm is developed which is named as nearest neighbour.	Pattern recognition is an application which is discovered by this algorithm. For example, a delivery boy delivering an online product to a house deems this algorithm very useful.
1979		Invention of Stanford cart	This car is moving in a room without any human effort when there is loss of huddles in room.
1981	Gerald Dejong	Technique of training the data set was introduced.	In this era computers are well set for removing such data which are not of use. To do this, computers make simple logic rules to train the data set.
1985	Terry Sejnowski and Charles Rosenberg	Discovery of NetTalk	Pronouncing letters is the main work done by NetTalk.
1986		Reshaping the neural network by some researchers.	Back propagation model is introduced in neural networks. By this technique, neural networks can create multiple layers to solve complex real-world problems.
1990		Introducing a method to learn the huge amount of data.	Many researchers are now able to develop such algorithms which can interpret the data of large quantity within less time as well as cost. This makes very easy to learn the thing in short time period.
1997		A computer-based company IBM has developed deep blue computer.	This computer is invented for playing of games especially chess. Ironically, a world champion of chess has been beat while playing the chess game on it.
1998		Research and development program initiated by AT&T Bell Laboratories.	Using back propagation technique for digital recognition.
2006	Geoffrey Hinton	A new development is taken up and is called deep learning (DL).	Algorithm which differentiates between objects and letters in various forms like videos as well as images.
2010	Alex Kipman	Microsoft introduces the interface Microsoft Kinect.	This device is developed on the bases of motion-sensing systems. In this interface, one can see the movement through designated systems. Gesture is also the main feature of this interface.

Year	Name	Description	Application
2011	Jeff Dean, Greg Corrado, Andrew Ng	IBM Watson is discovered by the IBM company. Another invention has introduced in real-world as Google brain.	IBM company has come up with great achievement. This interface is fully dedicated to AI. The main advantages of this device are reducing time, cost, and maximizing profit. It also provides evidence-based outcomes. In the same year Google brain is invented. It uses the concept of DL. Its functioning is very similar to that of the human brain. Its main application is based on patterns, its detection through photo capture. It is also applicable to YouTube videos to find the patterns.
2012		Invention of Google X Lab.	It is used for creating those algorithms which involve ML techniques for searching videos on YouTube.
2014		Deep face is introduced through Facebook.	A software has been created with the help of deep neural network (DNN) identify person while matching it with the photograph of a person. It has been seen the mechanism is same as that of human.
2015	Elon Musk	Introduced OpenAI company in the real world. Amazon introduced Amazon machine learning (AML) facility for various persons who are working in this field.	This company develops AI which helps mankind. It also provides solutions for unknown problems with AI. AML is basically an online platform which provides solutions to problems. It provides tools for visualizations by which projection can be done very easily.

4.4.2 Benefits of ML in the Healthcare Sector

In the current scenario the demand of improving the healthcare sector is the top priority of many countries, whether they are developed or developing ones. Bringing change in the healthcare sector is a very big task, but things can easily be managed by using those technologies which are built upon the ML techniques. ML contains such algorithms which are reliable,

TABLE 4.4   History of Healthcare with AI

Year	Name	Description	Application
1960	Dendral	A first system has developed which is based on AI to provide solutions for the given problem. It is known as Dendral.	Developers use it for organic chemistry.
1970	Jack D. Myers	In this year MYCIN has been introduced to the real world. This system has shown its role in making decisions about clinical reports. Also, MYCIN is part of Dendral.	This is the first system which is used in healthcare for providing results. This system detects infectious bacteria that can harm the human body. The medical practitioner can suggest antibiotics while checking the weight of the patient.
		Internist-1 starts functioning to provide solutions that occur in healthcare.	Basically, this system is a tool to provide evidence-based solutions while diagnosing the results.
1980		As the time changes rapidly, new approaches are introduced to healthcare systems.	Frameworks have been proposed in terms of ANN, Bayesian analysis, and fuzzy theory with AI technologies to enhance the healthcare system. These approaches provide solutions to many problems in healthcare.
1990		Concept of NN is introduced to healthcare sector.	With the help of computers, scientists use NN to upgrade healthcare systems like medicine. But NN didn't get the recognition at that time. The main reasons are the small data sets. Since NN is applicable to those large or complex data sets which provide result. NN techniques split the data set into two parts: train and test data sets.
2000 onwards		Different algorithms of ML are used to improve the healthcare sector. There are many subsectors of healthcare which increase working capabilities that make the system efficient as well as provide evidence-based solutions for complex problems.	The main areas which are highly influenced by ML are primary healthcare, shifting of manual work to digital work by maintaining the HER data, and enhancing the diagnose of disease. Monitoring different aliments with IoT-based devices, telemedicine, etc. Which can benefit the patients.

give accurate results, are cost efficient as well as provide evidence-based outcomes for making correct decisions in less time. These are very useful properties of ML to enhance the healthcare sector. Some of its benefits are described in Table 4.5.

TABLE 4.5   Benefits of ML in the Healthcare Sector

Name	ML Algorithms	Description
Identifying aliments and their diagnose		Algorithms based on ML are able to identify the cause of aliments. The main role is played by the huge amount of data that is recorded in the healthcare sector. ML algorithms provide accurate results for those data which are complex in nature. It gives answers to our query. For example, diseases like cancer are not easy to detect in initial stages. ML provides solutions to this kind of problem. If medical practitioners want to detect symptoms through omics data, it is very difficult to do it with conventional procedures. Again, ML give useful information when analysing omics information.
Disease prediction	Decision tree, Naive Bayes, KNN, clustering	Disease can be predicted very easily before it can harm the human body. ML has an ability to detect as well as predict the disease before initial stages. So those precautions can be taken well in time to increase a person's lifespan. For example, a very common disease is diabetes. This disease can affect various organs of the body if precautions are not taken in initial stages. Similarly, in the case of liver malfunction, ML algorithms provide best results in predicting this disease.
Image analysis		ML is becoming more reliable for analysing images to give accurate output. Companies like Microsoft are producing computers for the healthcare sector which analyse images, such as InnerEye.
Personalize healthcare		As a huge amount of data is collected in the field of healthcare sector, it is very difficult for medical practitioners to analyse the data. ML algorithms do a great job providing better solutions.

4.5 CASE STUDY

In this section, different case studies are considered, which show the importance of the healthcare sector.

4.5.1 AI Supporting Frontline Workers in Clinical Decision-Making

Non-adherence of maternal and child health-related guidelines at the field level by (ASHA and ANM) and non-compliance on post training follow-up in the field for these frontline workers has led to inconsistent clinical decision-making and poor referrals.

They meet patients in person either in their houses or during those days when the health sanitation and nutrition day is being held in that village. Often, they consult through telemedicine consultation with doctors at health and wellness centres. In the present situation, AI-based applications are developed in making clinical decisions in the healthcare sector. One such application is Arezzo by Elsevier. Such AIs is intended to help the frontline worker to make a clinical decision on his or her condition and suggest needed tests, medications, or referrals. By collating data from medical histories, symptoms, and medicines prescribed, the use of Arezzo-enabled medical guidelines can help various public health issues in the field.

Frontline workers in the healthcare sector face various problems in collecting patient data and maintaining patient records of patient vitals like blood pressure, heart rate, oxygen level, and other ailment when they reach out to rural areas where medical facility are on the edge. Some frontline workers have to make decisions for patients. Currently, new techniques are being developed by various companies on the basis of AI technologies. AI is helping frontline workers make clinical decisions about patient medications. In clinical practice, Arezzo excelling in making correct decisions. Some of the applications of Arezzo are given in Figure 4.7.

4.5.2 AI Support of Frontline Workers in Growth Monitoring of Newborn Babies

An NFHS 5 report depicts malnutrition and low birth weight as major ongoing public health issues. One of the process-based challenges is the availability of growth-monitoring tools at delivery points and Anganwadi centres.

A newborn baby with a weight of less than 2500 gm needs special care at the home-care level. Therefore, the accuracy of this weight is very crucial. It has been estimated that 20 million newborn babies are underweight. The main issue is in regularly monitoring the weight of the newborn for one

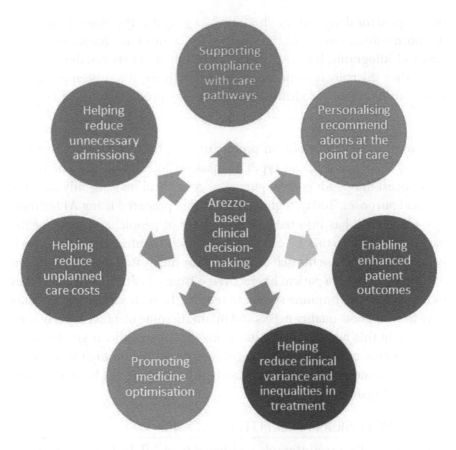

FIGURE 4.7   Functions of Arezzo by Elsevier.

to two weeks. It is not possible to go to the hospital every day for a mother who is coming from remote areas.

This issue again can be resolved with the help of AI-based techniques. Wadhwani AI is working in the same area to overcome these types of problems by providing AI-based solutions. Till now this group has made tremendous achievements in giving relief to mothers of newborn babies. They are developing a smartphone-based application with anthropometry technology. With this technique, frontline workers can easily click a 3D photograph using a smartphone and check for underweight babies in remote and rural areas. This machine helps frontline workers monitor newborn weight on a regular basis.

In the past it used to be difficult for frontline workers or doctors to know the weight of newborn babies. Most of the time, mothers had to travel to the hospital for daily routine check-ups, as weight is the major concern of newborn babies. Generally, the ideal weight of newborn babies should be 2.5 to 3.5 kilograms. If a newborn baby has weight in between the specified range then the baby is confirmed as a healthy baby. It has been estimated that 20 million newborn babies have to face the problem of underweight.

4.5.3 High-Tech Mask

Face masks are very useful in protecting people from various airborne viruses. There are different types of masks. Cloth masks are one of the commonest masks for general purposes. A medical mask is only used for medical purposes. Today, high-tech masks are prepared using AI technology. COVID-19 has infected a large number of people, and many of the patients died due to this disease. To control such an infectious disease, each person has to sanitise himself as well as wear medicated mask. It has been noticed that even if a patient has recovered from COVID -19 virus, it takes more days for the immune system to recover. In such a situation, patients have to wear good-quality masks and maintain some distance from others. As we fight this new virus, it is imperative that people wear good-quality masks to protect themselves. Scientists are designing on high-tech masks using AI technology, which may bring new concepts in the healthcare sector to fight against the deadly virus.

4.6 CONCLUSION AND FUTURE SCOPE

This chapter discussed the role and benefits of IoT in healthcare at different stages. IoT-based monitoring devices in the healthcare sector have been studied in detail. It has discussed how large amounts of information can be stored safely using the concept of big data in healthcare. This chapter also looked at ML in healthcare, including the history of ML as well as ML's roles and benefits in healthcare. Moreover, case studies were discussed to justify above techniques such as AI supporting frontline worker in clinical decision making, AI supporting frontline worker in growth monitoring of new born babies, and high-tech mask. All the above defined techniques have future scope to empower the healthcare sector which can deliver right solutions to the needy people. ML technique overcomes all the drawbacks in healthcare sector to make it more efficient as well as reliable. ML gains its remarkable achievement in different subsectors of healthcare which can benefit not only patients but also medical practitioners, pharmaceutical companies, insurance companies, and frontline workers.

REFERENCES

Aceto, G., Persico, V. & Pescapé, A. (2020). Industry 4.0 and health: Internet of things, big data, and cloud computing for healthcare 4.0, *Journal of Industrial Information Integration*, 18, 100129, https://doi.org/10.1016/j.jii.2020.100129.

Ajmera, P. & Jain, V. (2019). Modelling the barriers of health 4.0: The fourth healthcare industrial revolution in India by TISM, *Operations Management Research*, 12, pp. 129–145, https://doi.org/10.1007/s12063-019-00143-x

Farouk, A., Alahmadi, A., Ghose, S. & Mashatan, A. (2020). Blockchain platform for industrial healthcare: Vision and future opportunities, *Computer Communications*, 154, https://doi.org/10.1016/j.comcom.2020.02.058.

Hathaliya, J. J., Tanwar, S., Tyagi, S. & Kumar, N. (2019). Securing electronics healthcare records in healthcare 4.0: A biometric-based approach, *Computers & Electrical Engineering*, 76, pp. 398–410, https://doi.org/10.1016/j.compeleceng.2019.04.017.

Islam, Md. S., Hasan, Md. M., Wang, X., Germack, H. D. & Alam, Md. N. E. (2018). A systematic review on healthcare analytics: Application and theoretical perspective of data mining, *Healthcare*, 6, pp. 1–43.

Kadhim, K. T., Alsahlany, A. M., Wadi, S. M. & Kadhum, H. T. (2020). An overview of patient's health status monitoring system based on Internet of Things (IoT), *Wireless Personal Communications*, 114, pp. 2235–2262.

Kakria, P., Tripathi, N. K. & Kitipawang, P. (2015). A real-time health monitoring system for remote cardiac patients using smartphone and wearable sensors, *International Journal of Telemedicine and Applications*, pp. 1–11.

Koul, S. (2021). Machine learning and Deep learning in agriculture, in Patel et al. (Eds.), *Smart agriculture: Emerging pedagogies of deep learning machine learning and internet of things*, CRC Press, pp. 1–19.

Malasinghe, L. P., Ramzan, N. & Dahal, K. (2019). Remote patient monitoring: a comprehensive study, *Journal of Ambient Intelligence and Humanized Computing*, 10, pp. 57–76, https://doi.org/10.1007/s12652-017-0598-x.

Mehta, N., Pandit, A. & Shukla, S. (2019). Transforming healthcare with big data analytics and artificial intelligence: A systematic mapping study, *Journal of Biomedical Informatics*, 100, pp. 1–14.

Ojokoh, B. A., Samuel, O. W., Omisore, O. M., Sarumi, O. A., Idowu, P. A., Chimusa, E. R., Darwish, A., Adekoya, A. F. & Katsriku, F. A. (2020). Big data, analytics and artificial intelligence for sustainability, *Scientific African*, 9, pp. 1–5.

Saheb, T. & Izadi, L. (2019). Paradigm of IoT big data analytics in the healthcare industry: A review of scientific literature and mapping of research trends, *Telematics and Informatics*, 41, pp. 70–85.

Tan, E. T. & Halim, Z. A. (2019). Health care monitoring system and analytics based on internet of things framework, *ETE Journal of Research*, 65, pp. 653–660.

Tortorella, G. L., Fogliatto, F. S., Vergara, A. M. C., Vassolo, R. & Sawhney, R. (2020). Healthcare 4.0: Trends, challenges and research directions, *Production Planning & Control*, 31(15), pp. 1245–1260, DOI: 10.1080/09537287.2019.1702226.

Waschkau, A., Wilfling, D. & Steinhäuser, J. (2019). Are big data analytics helpful in caring for multimorbid patients in general practice?—A scoping review, *BMC Family Practice*, pp. 20–37.

Yin, Y., Zeng, Y., Chen, X. & Fan, Y. (2016). The internet of things in healthcare: An overview, *Journal of Industrial Information Integration*, 1, pp. 3–13.

CHAPTER 5

Real-Time IoT-Based Online Analysis to Improve Performance of PV Solar System for Medical Emergencies

Lokesh Varshney, Kanhaiya Kumar, Snigdha Sharma, and Dinesh Singh

CONTENTS

DOI: 10.1201/9781003227595-5

5.1 INTRODUCTION

This chapter focuses on the shading effect of different configurations used for enhancing the performance of photovoltaic (PV) systems and these systems work on real-time analysis with the use of IoT [1]. Shading spot adversely affect the efficiency of solar array which leads in increasing its cost. The drawbacks can be minimized using various configurations such as total cross tied (TCT) configuration, series-parallel (SP) configuration, Fibonacci series configuration, and SuDoKu configuration. This chapter highlights the five different configurations under different shading patterns. It has been observed that the SuDoKu power output configuration is better than other configurations.

The rapid development of sustainable energy is today demand for the development of nation because of their vital role in development of solar, wind, and much renewable energy. Photovoltaic (PV) is natural inhabitant and support natural environment and reduce the harmful fuel consumption. In India there are nearly 256 days of clear sunny sky per year so enormous amounts of solar energy are available, but it is limited because of low efficiency. PV performance efficiency is drastically affected by variations in sun irradiation, shading effects, and excess temperature rise of solar cells. These are the prominent reasons behind the deflection of solar panel maximum power point (MPP) that causes a reduction of overall efficiency. In this chapter, we describe how the system efficiency can be monitored continuously to improve the output by using IoT and vision sensors [2].

Srinivasan et al. [3] designed a solar monitoring system using Current Sensor ASC712, voltage sensor, heat sensor, Temperature and Humidity Sensor RHT03, Adafruit Cloud and Arduino Mega 2560. By using this hardware, the different parameters of solar energy are measured and analysed so steps can be taken to improve system efficiency. This system efficiency improved by 95 percent. The data kept in cloud memory can also be analysed by MathWorks®. Cloud memory provides a CSV for examination of R language.

Youssef Cheddadi et al. [4] developed low-cost IoT-based sun power nursing system. The system finds the electrical and environmental parameters of a PV system. Smart sensors, controllers, and algorithms for solar nursing modules are applied to grab real-time data and supplied to user. This decreases the effective costs of measurement instruments.

Lokesh Babu R.L R et al. [5] discussed IoT-based dust monitoring system in which solar panels get dust removed continuously to maximize the output energy, and an active operation of solar panel is maintained. The module also confirms failures of the PV system and whether electrical utility is performing directly on solar module or loads. Entire PV modules are coupled, and sensors are allied to main control units which govern PV and loads. By integrating IoT equipment, the information of PV parameter and data through cloud memory will continuously screen to remote subscriber. Users can view present, prior, and average readings of PV parameters like intensity of light, current, voltage, and temperature via graphical representation.

Andreas [6] developed a remote analysis and control of PV isolated solar farming for smart default identification and energy optimization. The administration of a PV string needs supplementary electrics and electronics that are involved in each solar PV panel. The collaboration of industry and university government project development was recognised for making an intelligent nursing device and founding allied algorithms and software for default identification and solar module administration. The latest smart monitoring device contains a radio and relays which allow users to adjust PV module linking topologies. This PV-powered IoT module is presently programmable and provide a) data analysis through mobile, b) control of PV farms, c) identification of faults and their rectification, d) optimization on different shading power, and e) the developed inverter transient is minimized. This project describes PV module efficiency and become simple with machine learning.

Vinay Gupta et al. [7] designed an economical, reliable, and affordable DAQ (data acquisition) system for sensing and nursing of the photovoltaic system. This problem can be resolved by applying IoT-based DAQ systems. The established DAQ system works on public access cloud memory and software. This project presents a design of a cost-effective DAQ system in lieu of collecting operational data of the PV system for examination of system efficiency.

Xiangyun Qing et al. [8] proposed a model for the improvement in the features of SP configuration under various shading situations, varying irradiance, and temperature variants. Yuki Mochizuki et al. [9] designed a system which produces power 1.5 times that of flat surface PV panel which is equal to dual axis panel. Fibonacci number PV modules include many features; for instance, reflected light from one cell can be used by other cells, stages can be enlarged, which has minor spot with individual shadow.

Mochizuki and Yachi [10] presented the Fibonacci technique grounded on honey comb configuration to achieve optimum features of power production. According to the discussed design, a Fibonacci-centred solar PV system is placed about 30 centimetres away from the centre of the solar PV module.

To enhance the output power from the solar PV system, solar modules are connected in various configurations. Less efficiency is the effect of mismatch losses in the solar PV system caused by partial shading. Within shading situations, series structure starts behaving like a load and not as a source. So various configurations are developed to diminish the losses which occur because of shading.

5.2 SHADING CONFIGURATIONS

Less immersion of shading can also critically affect solar PV module efficiency. For output examination, different arrangements are implemented in four shading arrangements. These arrangements are reliant on the number of shading columns and total shading modules on each column. The types of arrangements are short wide (SW), long wide (LW), short narrow (SN), long narrow (LN) [11], as shown in Figure 5.1. Diagonal arrangement is also considered, and all arrangements are implemented on 5 x 5 PV array configuration.

5.3 VARIOUS ARRANGEMENTS BEHAVIOUR UNDER DIVERSE SHADING ARRANGEMENTS

5.3.1 Series-Parallel (SP)

Series-parallel combinations are used to enhance the overall output power of solar PV systems, as shown in Figure 5.2. Larger numbers of series-connected modules provide greater voltage, and larger numbers of parallel-connected modules provides greater current [12]–[16].

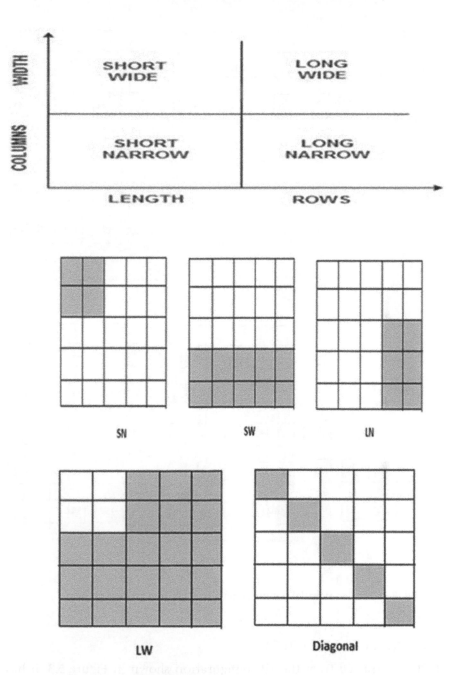

FIGURE 5.1   Types of shading patterns.

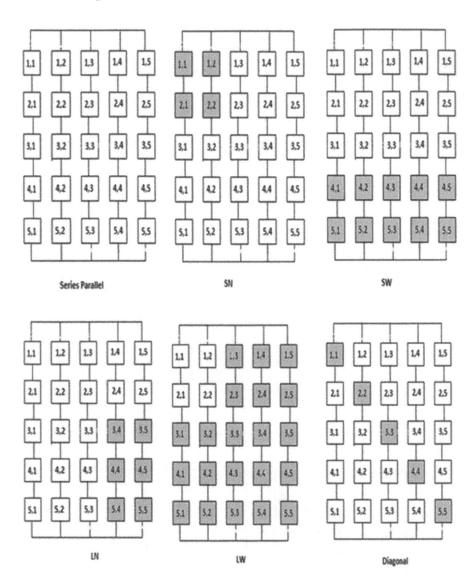

FIGURE 5.2   Series-parallel connections

5.3.2 Total Cross Tied (TCT)

TCT is extracted from the SP configuration shown in Figure 5.3. It has some additive characteristics of angular linked columns and rows such that the entire voltage and current values are identical through all columns and rows respectively. It works better than SP in shading conditions, but it increases cable losses as the numbers of ties increase.

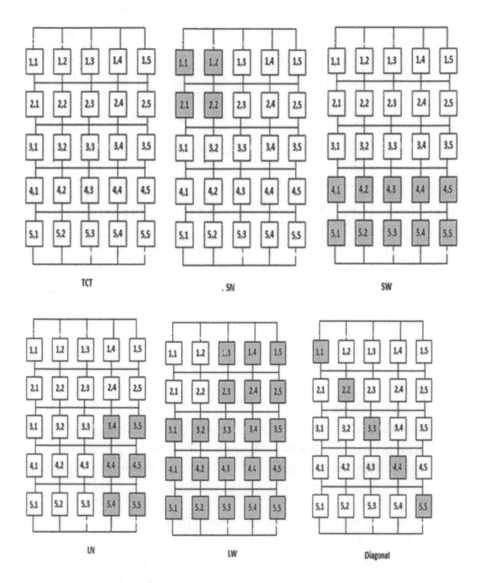

FIGURE 5.3   TCT connection.

5.3.2.1 Bridge Link

The bridge configuration is the root of bridge link (BL) configuration, as shown in Figure 5.4. It is attained from TCT with the benefit of a small number of ties and low cable losses but, it has a demerit that it disturbs complete the current and voltage under shading situations.

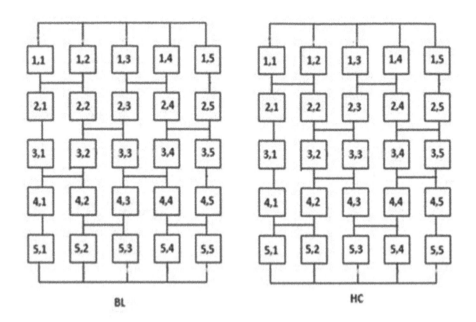

FIGURE 5.4   BL and HC connections.

5.3.2.2 Honey Comb

Honey comb structure is also derived from TCT based on honey comb (HC) arrangement, as shown in Figure 5.4. The output power losses can be minimized in this configuration but it has a restriction that it cannot decrease power losses under all shading conditions.

5.3.2.3 Physical Relocation of Module with Fixed Electrical Connections (PRM-FEC)

This pattern has the benefit of scattering of shaded portions to ignore more shading in the same column or row. In this technique, electrical internetworks remain unchanged with the rearrangement of the solar PV module, as shown in Figure 5.5.

5.3.2.4 SuDoKu

A 5 × 5 TCT matrix is reflected in the SuDoKu arrangement whose output is better within shading situations, as shown in Figure 5.6. In this arrangement, module rearrangement arises unrelated to electrical networks. Physical rearrangement of the modules leads in enhancing the length and price of wires.

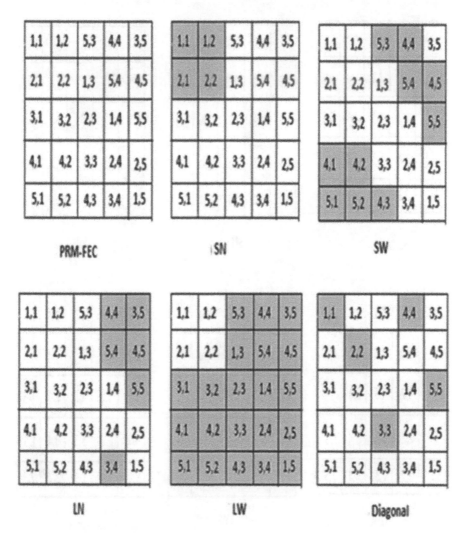

FIGURE 5.5   PRM-FEC pattern.

5.3.2.5 Fibonacci Series

In this arrangement, the solar module replaces leaves on a stem to boost the output efficiency of the solar PV panel. The major cause of losses in this configuration is shading by higher panels on lower panels. The Fibonacci pattern phyllotaxis is considered 1/3, 2/5 and so on, where phyllotaxis numerator gives leaf turns number and its denominator represents leaf number [17–20].

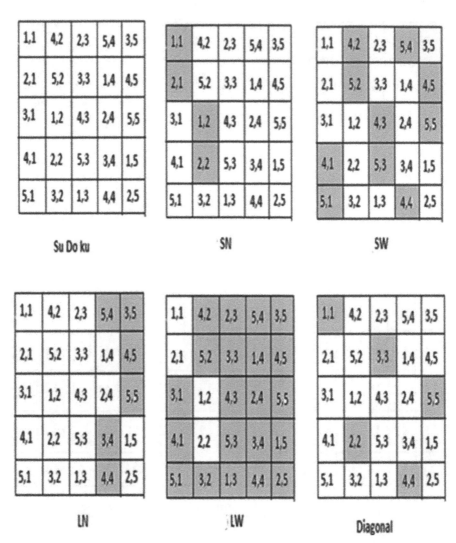

FIGURE 5.6 SuDoKu pattern.

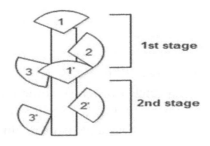

FIGURE 5.7 Arrangement of Fibonacci modules.

5.4 PV MODULE AND COMPARISON BETWEEN DIFFERENT CONFIGURATIONS

The PV array has five units in a series forming a string while all strings are connected in parallel. In the 5 × 5 matrix, a total of 25 modules are connected. The array has a nearly 4.5 kW capacity. Specifications of a PV module at standard test conditions are shown in Table 5.1.

All the configurations are analysed for various shading patterns, which permit us to identify the best configuration for particular applications. Table 5.2 and Figure 5.8 establish data about different configurations [21–24].

5.5 REAL-TIME IOT CONFIGURATIONS

The different configurations included in this chapter were analysed online using MATLAB® tools. All individual configurations are carried out on MATLAB tools which integrate their data online in order to monitor the performance of different parameters of the PV panel continuously. Here, the solar-panel-generated DC power is given to Arduino to fetch current,

TABLE 5.1 PV Module Specifications

PV Panel	TP 180
Panel size	1587 mm × 790 mm × 50 mm
Panel weight	16 kg
P_{max}	180 W
V_{MPP}	35.8 V
I_{MPP}	5.03 A
V_{oc}	43.6 V
I_{SC}	5.48 A

TABLE 5.2 Comparison between Different Configurations

Shading Patterns	SP	TCT	PRM-FEC	SuDoKu
	P (kW)	P (kW)	P (kW)	P (kW)
SN	3.51	3.57	3.58	4.11
SW	2.89	2.94	3.10	3.14
LN	3.40	3.46	3.51	3.60
LW	2.78	2.85	2.95	3.07
Diagonal	3.95	4.01	4.07	4.12

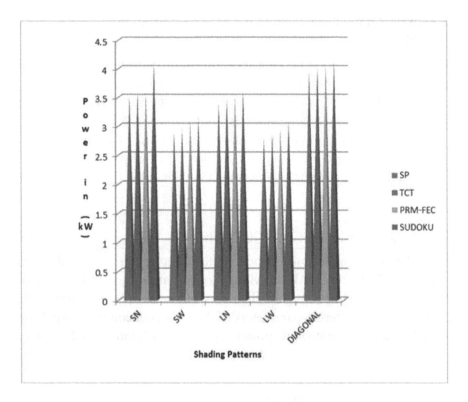

FIGURE 5.8   Power generated under different shading patterns.

voltage, power values, as shown in Figure 5.9. The calculated value of the solar parameters is given to a Raspberry Pi 4B with 8 GB RAM and a high-speed processor and is transmitted through cloud computing to users continuously analysing the data on real time, as shown in Figure 5.10.

5.6 APPLICATIONS IN HEALTHCARE

This model can be used in healthcare in many ways. Mobile clinics and ambulances can be equipped or configured with lifesaving equipment which is supplied power by solar panel installed on the roof of the vehicle. IoT-based systems will also monitor and transmit vital parameters such as heart rate and blood pressure through the internet to medical experts. These vital parameters provided by the devices can be used for analysis by health experts without physically observing the patient. These devices which can be easily worn by the patient can even alert health experts when patients are experiencing arrhythmias, palpitations, strokes, or heart attacks. Optimal

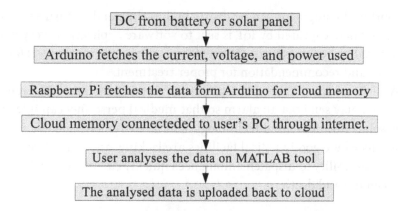

FIGURE 5.9   Workflow of real-time IoT system.

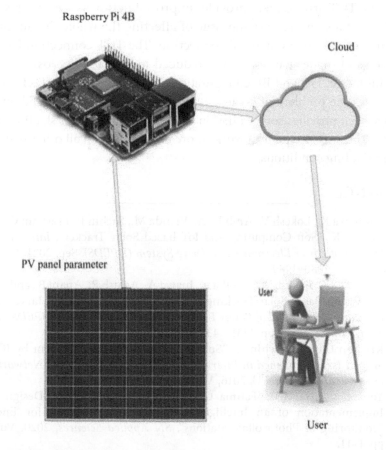

FIGURE 5.10   Real-time configuration.

support and equipment can be dispatched in time and in turn save many lives. The data captured by IoT is sent to software applications very closely monitored by health experts. Different types of algorithms are used for data analysis and recommendation for proper treatment.

An IoT sensor that detects a patient's exceptionally low heart rate, for example, may send out an alarm so that medical personnel can intervene. This solar- and IoT-based technology will be extremely valuable in rural locations where good medical facilities are lacking, and many lives will be saved. The ability to dispatch ambulances quickly can mean the difference between life and death.

5.7 CONCLUSION

This real-time IoT-based analysis finds that SP connection has merits such as simple implementation, non-redundant connections, and less wiring time. The TCT arrangement provides improved performance with respect to SP and it also overcomes the issue of affecting the whole thread under shading circumstances in an SP connection. The TCT connection has the advantage of higher harvest power, reduced incompatibility losses, and a large fill factor. HC and BL configurations are obtained from TCT which overcomes the problem of greater numbers of ties in TCT. The Fibonacci arrangement provides solar radiation to every module under shading conditions. The SuDoKu pattern works more efficiently than all other patterns during shading conditions.

REFERENCE

[1] Kanhaiya K., Lokesh V., Ambika A., Vrinda M., Sachin P., Prashant C., and Namya K. "Soft Computing and IoT based Solar Tracker", *International Journal of Power Electronics and Drive System (IJPEDS)*, Sep. 2021, Vol. 12, No. 3, pp. 1880–1889.

[2] Kanhaiya K., Lokesh V., Ambika A., Inayat A., Ashish R., Anant B., and Sajal. O. "Vision Based Solar Tracking System for Efficient Energy Harvesting", *International Journal of Power Electronics and Drive System (IJPEDS)*, Sep. 2021, Vol. 12, No. 3, pp. 1431–1438.

[3] Krishan G., and Sachin C. "Solar Energy Monitoring System by IOT", *Special Issue Published in International Journal of Advanced Networking & Applications (IJANA)*, 2016, Vol. 7, No. 4, pp. 46–51.

[4] Youssef C., Hafsa C., Fatima C., Fatima E., and Najia E. "Design and Implementation of an Intelligent Low-Cost IoT Solution for Energy Monitoring of Photovoltaic Stations", *SN Applied Sciences*, 2020, Vol. 2, pp. 1–11.

[5] R. L. R. Lokesh Babu, D. Rambabu, A. Rajesh Naidu, R. D. Prasad, and Gopi, K. "IoT Enabled Solar Power Monitoring System", *International Journal of Engineering & Technology*, 2018, Vol. 7, No. 3.12, pp. 526–530.

[6] Andreas S. S. "Solar Energy Management as an Internet of Things (IoT) Application", 2017 8th International Conference on Information, Intelligence, Systems & Applications (IISA), IEEE, 2017, pp. 1–4.

[7] Vinay G., Madhu S., Rupendra K. P., and K. N. Dinesh Babu. "A Low-Cost Real-Time IoT Enabled Data Acquisition System for Monitoring of PV System", *Energy Sources, Part A: Recovery, Utilization, and Environmental Effects*, 2020, pp. 1–16.

[8] Xiangyun Q., Hao S., Xiangsai F., and C. Y. Chung. "Submodule-Based Modeling and Simulation of a Series-Parallel Photovoltaic Array under Mismatch Conditions", *IEEE Journal of Photovoltaics*, 2017, Vol. 7, pp. 1–9, doi: 10.1109/JPHOTOV.2017.2746265.

[9] Yuki M., and Toshaiki Y., "Relationship between Power Generated and Series/Parallel Solar Panel Configurations for 3D Fibonacci PV Modules", 6th International Conference on Renewable Energy Research and Applications (ICRERA), San Diego, CA, USA, Nov. 2017, pp. 126–130.

[10] Yuki M., and Toshaiki Y. "Effective Series-Parallel Cell Configuration in Solar Panels for FPM Power Generation Forest", 7th International Conference on Renewable Energy Research and Applications (ICRERA), Paris, France, Oct. 2018, pp. 294–300.

[11] Darmini V., and Sunitha K. "Comparison of Solar PV Array Configuration Methods under Different Shading Patterns", IEEE International Conference on Technological Advancements in Power and Energy (TAP Energy), Kollam, India, Dec. 2017.

[12] Snigdha S., Lokesh V., Rajvikram M. E., Akanksha S., Aanchal S., R. K. Saket, Umashankar S., and Eklas H. "Performance Enhancement of PV System Configurations under Partial Shading Conditions Using MS Method", IEEE Access, 2021, G S I T S, Indore, MP.

[13] Snigdha S., Manasi P., Meenakshi S., and Lokesh V. "Comprehension of Different Techniques used in Increasing Output of Photovoltaic System," Presented in International Conference on Electrical and Electronics Engineering (ICEEE 2020) SPRINGER, NPTI, Faridabad, 28th–29th Feb. 2020.

[14] Ahmed A. M., Md. Ruhul Amin, and Kazi Khairul I. "Determination of Module Rearrangement Techniques for Non-Uniformly Aged PV Arrays with SP, TCT, BL and HC Configurations for Maximum Power Output", International Conference on Electrical, Computer and Communication Engineering (ECCE), Cox'sBazar, Bangladesh, Feb. 2019.

[15] Lokesh V., and Shivi R. "Enhanced Power Generation from Piezoelectric System under Partial Vibration Condition", The proceeding of IEEE International Women in Engineering (WIE) Conference on Electrical and Computer Engineering (WIECON-ECE) at Women Institute of Technology, Dehradun, Uttarkhand, India, 18th–19th Dec. 2017.

[16] Lokesh V., and Ambesh P. U. "Comparison of Techniques for Designing and Modeling of High Power Piezoelectric Devices", The proceeding of 4th IEEE Uttar Pradesh Section International Conference on Electrical, Computer and Electronics (UPCON), Mathura, India, 2017.

[17] Balaji V., Takaharu T., Aberra J., and Tefera M. "Mismatch Loss Analysis of PV ArrayConfigurations under Partial Shading Conditions", 7th International Conference on Renewable Energy Research and Applications (ICRERA), Paris, France, Oct. 2018, pp. 1162–1167.

[18] Shivi R., Lokesh V., A. Mishra, and S. Joshi. "Electric Power Generation from Piezoelectric System under Several Configurations", The proceeding of International IEEE Conference on Computing, Power and Communication Technologies 2018 (GUCON), 28th–29th Sep. 2018.

[19] S. Vijayalekshmy, G. R. Bindu, and S. Rama Iyer. "Performance Comparison of Zig-Zag and Su Do Ku Schemes in a Partially Shaded Photo Voltaic Array under Static Shadow Conditions", International Conference on Innovations in Power and Advanced Computing Technologies (i-PACT), Vellore, India, 2017, pp. 1–6.

[20] Ritu S., Rakesh Y., Snigdha S., and Lokesh V. "Analysis and Comparison of PV Array MPPT Techniques to Increase Output Power", Presented in International Conference on Advance Computing and Innovative Technologies in Engineering (IEEE), 4th & 5th Mar. 2021.

[21] Kanhaiya K., Tanya A., Vishal V., Suraj S., Shivendra S., and Lokesh V. "Modeling and Simulation of Hybrid System", *International Journal of Advanced Science and Technology*, 2020, Vol. 29, No. 4, pp. 2857–2867, http://sersc.org/journals/index.php/IJAST/article/view/22180.

[22] Ahmad Sohrab A., and Lokesh V. "Performance Improvement of Solar PV under Partial Shading Conditions", Presented in International Conference on Intelligent Technologies (CONIT *IEEE*-2021), Karnataka, India, 25th–27th June 2021.

[23] Dinesh S., Junaid A., Sameer A., and Lokesh V. "Performance Analysis of Footstep Power Generation using Piezoelectric Sensors", Presented in International Conference on Intelligent Technologies (CONIT *IEEE*-2021), Karnataka, India, 25th–27th June, 2021.

[24] Kanhaiya K., Lokesh V., Aladiyan A., R. K. Saket, and S. Mekhilef. "Solar Tracker Transcript: A Review", *International Transactions on Electrical Energy Systems*, 2021, e13250, http://doi.org./10.1002/2050-7038.13250.

CHAPTER 6

Major Depressive Disorder Detection and Monitoring Using Smart Wearable Devices with Multi-Feature Sensing

Shamla Mantri, Seema Nayak, Ritom Gupta, Pranav Bakre, Pratik Gorade, and Vignesh Iyer

CONTENTS

DOI: 10.1201/9781003227595-6

6.1 INTRODUCTION

Mental health disorders, more specifically major depressive disorder (MDD), is one of the leading worldwide health concerns. Patients often report mood swings, personality issues, inability to cope up with seemingly normal problems, frequent stress, withdrawal from social activities and friends, etc. In 2010, mental health problems were the leading causes of years lived with disability (YLD) worldwide. Anxiety and depressive disorders were among the most frequent disorders [1]. In recent times, there has been growing appreciation of the dynamic nature of a depressive symptom, with several studies suggesting the clinical significance of temporal fluctuations of symptoms, including suicidal thoughts. The chronic and relapsing nature of several mental health disorders are the norm, and not a quirk, hence patients are required to undergo long-term follow up, and assessment methods of several kinds become essential for understanding symptoms as well as recovery methods. However, most clinical assessment tools for MDD symptoms need to be administered at certain fixed intervals and time points and are subject to retrospective recall. Evaluation methods harnessed by psychologists are prone to impressionistic inclinations. An alternative proposed is the Ecological Momentary Assessment (EMA) which complements scales with static rating by providing almost real-time symptom measurement and regular monitoring of thoughts, feelings, and behaviours. It is our strong conviction that conducting the same on users' smartphones will overcome bias due to the Hawthorne effect of traditional laboratory or clinical set-ups. Developing applications which can be used individually, are clinically applicable and feasible, and can be scaled using modern-day technologies will be the leading change in making such solutions truly ubiquitous.

This work is a proposed solution for monitoring MDD using data collected from smart wearable devices such as fitness trackers and responses to the PHQ-9 questionnaire. With consistent data collection and use of machine learning techniques we propose a solution that helps in classifying patients into depressed and control categories.

Such a mechanism system requires patients to own and use a smart band with heart rate and sleep tracking capabilities for monitoring and a computer or mobile device to fill out the questionnaire. All this data is processed by models to classify the subject. To reduce the inherent complexity due to the nature of the solution, we chose to stick to what can be captured by even the most basic smart wearable devices. Many previous works have espoused heart rate variability (HRV) [2] and sleep patterns [3,4] for their noticeable correlation with depressive disorders. Both can be easily measured using any consumer-level smartwatch or fitness tracker available on the market; however, they function with varying levels of accuracy.

We developed a web application which tacitly can run across several browsers including Google Chrome, Safari, and Firefox. Additionally, it can also be used as a native app by virtue of its progressive web application (PWA) features, so users can also install it on their devices without having to continuously remember and search for the URL. The application is a portal to collect response and health measurement data from users at varying periods. It can thus assess a patient's state of mind, thrice a day, from a randomized subset of three questions from the DSM-5 standardized PHQ-9 questionnaire, chosen for its cogency across various platforms. A browser-based implementation helps us in easing the process of data capturing across differing web and native app platforms, along with enabling us to collect variability of symptoms across days, and accurately represent intraindividual symptoms and their variation.

6.2 LITERATURE SURVEY

6.2.1 PHQ-9

PHQ-9 is a DSM-5 approved mental health questionnaire consisting of nine questions answered by the subject. A PHQ-9 questionnaire has a sensitivity of 88 percent and specificity of 88 percent for PHQ Scores ≥ 10, implying PHQ-9 is a reliable measure for identifying moderate and severe levels of depression [5]. The severity measures for PHQ scores are shown in Table 6.1.

6.2.2 Heart Rate Variability

Heart rate variability (HRV) is one of the metrics which can be directly captured or can be calculated from the heartbeat of the subject (from the wearable device). It is defined as the variation between heartbeats over

TABLE 6.1 PHQ-9 Severity Scores

PHQ-9 score	Severity
0–4	None
5–9	Mild
10–14	Moderate
15–19	Moderately Severe
20–27	Severe

a period. Notably, this metric does not depend on heart rate at a given instance, but on how the heart rate fluctuates. Generally, higher HRV is associated with good health and low HRV with illness [6]. Before antidepressant treatment, patients with MDD show reduced HRV compared to healthy patients [7].

In psychological studies, researchers have found HRV being directly proportional to better self-control [8], social skills [9], and higher abilities to cope with stress [10]. A similarity between these is their direct or indirect relation to mood and, hence, depression.

HRV has a wide range of parameters, which can be broadly divided into three categories. Time domain (RMSSD, SDNN, TINN), frequency domain (LF, HF), and non-linear (SD1 and SD2) [11]. In all three categories, research has indicated a significant correlation between HRV and MDD [12,13]. Recent studies stipulate non-linear parameters having better accuracy over the linear parameters [12].

6.2.3 Sleep

Sleep woes have been one of the key indicators of depressive disorder. In many clinical cases, sleep-related trouble has been the primary cause for depressed patients to seek help. Seventy-five percent of depressed patients show signs of insomnia, with 40 percent of young adults suffering from depression showing signs of hypersomnia [14]. Furthermore, there has been evidence of sleep pattern being related to mood [3] and abnormal sleep pattern in depression [4]. Antidepressants show some success in altering sleep patterns towards normalcy [15]. This establishes insomnia as an effect of MDD.

Studies have found insomnia as a cause of depression. In multiple studies, patients interviewed who revealed insomniac conditions had been at a greater risk of being depressed after a few years [16].

In a Dartmouth University study, students' sleep was detected using automatic sensor data from smartphone to confirm the relation between the inability to sleep and depression. In sync with other findings, there is a negative correlation between sleep and depression. This indicated students who slept less were more likely to suffer depression [17,18].

Another approach to predict depression from sleep is from the sleep pattern. The different phases of sleep give a rough idea about what could be a normal sleep pattern. Significant differences can be exploited to classify patients with depression symptoms. Sleep is classified into two phases: non-rapid eye movement (NREM) and rapid eye movement (REM) sleep.

Abnormalities in sleep of depressed patients are mainly indicated by increased REM sleep and reduced slow wave sleep [19]. Reduced REM latency can also be used to predict depression [20].

Currently, the sleep data from smartwatches is derived using accelerometers and heart rate monitors. They sense movement during sleep and differentiate between light and deep sleep based on heart rate. Based on the derived statistics, they calculate a sleep score.

6.3 METHODOLOGY

6.3.1 System Design/Procedure

Users are initially required to submit their baseline activity data including heart rate, sleep, and responses to questions presented in a web application. A new account registered in the application will be approved by an existing administrator before it can be used to upload any data to the datastore. This is to ensure the service is not misused and remains generally available till the stipulated date. Any device with an internet-connected browser with a minimum connection speed of 500 Kbps can be used to access the service. Exclusion criteria for using the application include a user's refusal to participate or their clinician's concerns that participation might prove to be harmful in any manner.

At the onset, users are required to fill in the standard e-version of the traditionally paper-based PHQ-9 survey to obtain an initial understanding of their mental condition. The application alerts users whenever a new question set is ready to be answered, which is typically scheduled three times in a day: 9 a.m.–12 p.m. (after waking up), 2–5 p.m. (awake), 7–10 p.m. (before sleep). Responses are expected within a three-hour timeframe to ensure that the responses reflect a very recent or current mental conditions. In each question set, users are asked to respond to three questions from the PHQ-9 questionnaire, sampled from the standard set of nine questions, without replacement, and a Likert scale is used to obtain scaled responses. As a result, some questions could be asked twice or more times per day, while some, not at all.

To avoid survey fatigue for the user, as well as to ensure responses are not biased towards the phrasing of any query, each question can be presented in two variants, one where users can mark if they *are feeling depressed* or *not feeling depressed* during different instances of the day, both measured on opposite scales of 0–3, 0 being the minimum and 3 being the maximum value.

Users are required to continuously wear a fitness tracker equipped with a photoplethysmography (PPG) sensor, designed to measure heart rate by evaluating skin perfusion based on refraction and absorption of mid-wavelength visible (i.e., green) light, which it can detect. This method does suffer from inaccurate measurement as compared to gold-standard measures like electrocardiography, and our application is by nature as accurate as the readings obtained from such a sensor. Users can upload their health data including their sleep and heart rate, which is then available to export from the OEM host app, subject to in-app feature availability of exporting the aforementioned health data.

Since these apps sometimes restrict functionality, including heart rate measurement frequency, it is also necessary to pair a device with an open-source replacement host application like Gadgetbridge, by means of which a custom duration and period of sensor activation can be set. Users can also upload their heart rate values from the SQLite database generated by Gadgetbridge, to obtain a more accurate insight into HRV values.

A typical period of relevant data collection is set as a rolling window of 14 days, excluding the days where no data is available. Using the heart rate data, a standard HRV analysis is conducted, whereby two HRV features are considered to have the highest order of median rank from the classifications based on different sets of input data, namely, SD2 and SDNN [12]. SD1 and SD2 from the Poincare plot [21] can be derived by combining some time-domain features, RMSSD and SDNN, though SD1 and SD2 [22] can be directly obtained from the raw data.

The application can also be made to function as a depression indicator by performing a set of activities. The set of activities is to be performed by the user wearing the band. This whole process takes around 30 minutes, then the data has to be extracted from the host application and uploaded. There has to be a gap of 30 seconds to one minute This will give a measure of depression, which however, is not as accurate as the one measured with data over a period of time.

6.3.2 Heart Rate

We used the SWELL dataset [23] with SD1 and SD2 values filtered out and some tweaks to make it a dataset for depression detection. This dataset was used to train the models. HRV features used in the presented solution are listed in Table 6.2. We rendered Poincare plots which are a visual

TABLE 6.2   Heart Rate Parameters

Domain	Parameter	Unit	Description
Poincare plot	SD1	milliseconds	Standard deviation of the Poincare plot perpendicular to (SD1) and along (SD2) the line of identity
	SD2	milliseconds	

representation of SD1 and SD2 values. The Poincare plots between RR_n and RR_{n+1} is shown in Figure 6.1 for all phases.

6.3.3 Sleep

The data for sleep is captured by the band using its accelerometer and heart rate monitor. Based on heart rate readings it classifies sleep into light and deep sleep. This data can be captured from the host app (MiFit, Huawei Health, etc.). We used this data from the host app to assign a sleep score on a Likert scale. This score was further adjusted with the PHQ-9 scores for the questions specific to sleep. In the event of the user not answering any of the PHQ-9 questions, the sleep score takes that place. Hence, the sleep score value will adjust the PHQ score, further validating the PHQ-9.

There can be different conditions in such a scenario. Both sleep data and user PHQ input are given, only sleep data is given, only PHQ data is given, or none of them is given. There is adequate handling for the first three conditions.

6.3.4 Classification and Performance Evaluation

We used three classifiers to categorize a pair of SD1 and SD2 values into either depressed or not depressed. Based on accuracy, execution time, and computational load, we selected the classifier to be used in our software. The performance of classifiers in terms of accuracy and execution time is shown in Table 6.3.

Even though, the Random Forest Classifier gave a greater accuracy than the AdaBoost-decision tree combination, its execution time was higher and the load on the computer was greater. The error rate graphs for KNN and AdaBoost classifiers are shown in Figure 6.2.

Hence, we chose the AdaBoost-decision tree combination. The confusion matrix for AdaBoost classifier is shown in Figure 6.3.

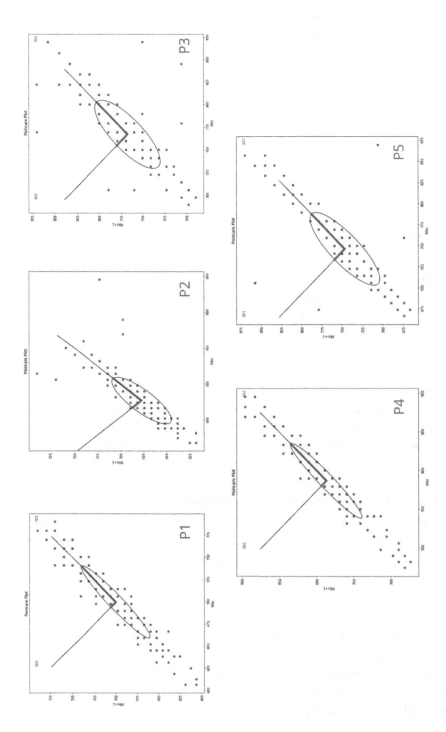

FIGURE 6.1   Error rate of AdaBoost classifier for different value of n_estimators ranging from 50–100.

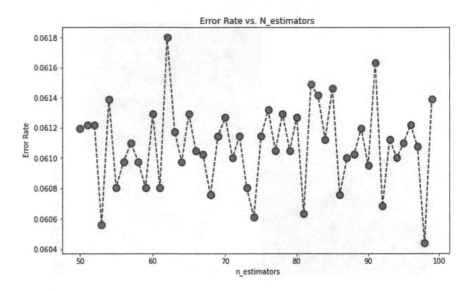

FIGURE 6.2A   Error rate of AdaBoost classifier for different values of n_estimators ranging from 50–100.

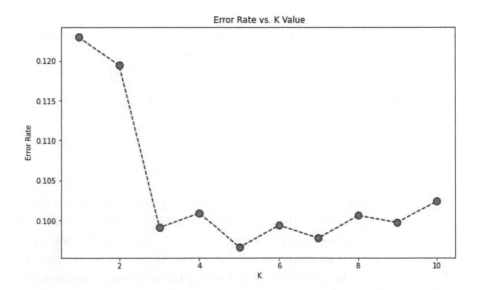

FIGURE 6.2B   Error rate of KNN classifier for different value of K ranging from 1–10.

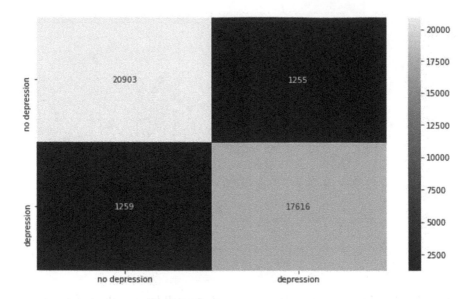

FIGURE 6.3   Confusion matrix of AdaBoost showing classification results.

TABLE 6.3   Performance of KNN, Random Forest, and AdaBoost Classifier for Our Dataset.

Classifier	Training Time	Parameters	Accuracy
KNN	4s	n_neighbours = 5	90.3%
Random forest	1 m 16 s	N_estimators = 100, max_depth = None (Default)	94.9%
AdaBoost classifier with decision tree as base classifier	4s	N_estimators = 97, learning rate = 1,	93.34%

6.4 RESULTS

We got an accuracy of 93.3 percent in detecting depressed mood from HRV using the decision tree and AdaBoost classifier combination. We have developed a system which can perform consistent mood monitoring using sleep, HRV, and PHQ-9 scores. Our solution uses only non-linear HRV parameters. In previous works, linear parameters or a combination of linear and non-linear parameters were used [12].

A comparative analysis of our results with a solution comprising classification based on linear and non-linear HRV parameter is shown in Table 6.4.

TABLE 6.4   Comparative Analysis between Solutions in [12] and Our Solution

Parameter	Solution in [12]	Our solution
Sensitivity	73%	93.39%
Specificity	75.6%	94.33%
Accuracy	74.4%	93.34%

6.5 LIMITATIONS

The data used in this work is data captured by wearable sensors. Most experiments in this field have been done using ECG signals, and the accuracy of bands in detecting heart rate is unlikely to be at par with such advanced systems. Nevertheless, our implementation can work with ECG data. Owing to COVID-19 restrictions we could not obtain too many real samples, which we can fix in future research.

6.6 CONCLUSION AND FUTURE WORK

Using data collected from the smart wearable devices and answers to PHQ-9 questionnaires by the user, we developed a mechanism to detect and measure the severity of depression in a subject. We achieved an accuracy of 93.3 percent using the dataset available to us.

These results indicate the possibility of detecting depression with decent accuracy by using only a small number of parameters. With more accurate sensors in the future, and the inclusion of related parameters like oxygen saturation (SpO2) [22] in smart bands, the accuracy of our solution should increase. The usage of low-cost, low-power smart bands and fitness trackers as a full-fledged solution for mental health diagnosis is not far away.

The consistent surge in adoption of bands among the younger population (the group most affected by mental health disorders [24]) will further bolster the use of these technologies to play a key role in healthcare.

6.7 ACKNOWLEDGEMENTS

We wish to express our sincere gratitude to Prof. Dr. Shamla Mantri as well as Prof. Dr. Vrushali Kulkarni, H.O.S, School of CET, MIT World Peace University, Pune, for guiding us in this survey. We also thank our friends and other MIT World Peace University staff members for guidance and encouragement in carrying out this work, and for their support.

REFERENCES

[1] Torous J. et al. "Utilizing a Personal Smartphone Custom App to Assess the Patient Health Questionnaire-9 (PHQ-9) Depressive Symptoms in Patients With Major Depressive Disorder." *JMIR mental health* vol. 2,1 e8. 24 Mar. 2015, doi:10.2196/mental.3889

[2] Sangwon B. Young K. A., Hye J. E. et al. Detection of major depressive disorder from linear and nonlinear heart rate variability features during mental task protocol, Computers in Biology and Medicine, vol. 112, 2019, 103381, ISSN 0010-4825, https://doi.org/10.1016/j.compbiomed.2019.103381.

[3] Kroenke K., Spitzer R. L and Williams J. B. "The PHQ-9: validity of a brief depression severity measure." *Journal of general internal medicine* vol. 16,9 (2001): 606-13. doi:10.1046/j.1525-1497.2001.016009606.x

[4] Vaccarino V., Lampert R., Bremner J. D. et al. "Depressive symptoms and heart rate variability: evidence for a shared genetic substrate in a study of twins." *Psychosomatic medicine* vol. 70,6 (2008): 628-36. doi:10.1097/PSY.0b013e31817bcc9e

[5] Dekker J. M., Crow R. S., Folsom A. R. et al. Low Heart Rate Variability in a 2-Minute Rhythm Strip Predicts Risk of Coronary Heart Disease and Mortality From Several Causes, *Circulation*, vol. 102,11 (2000): 1239–1244

[6] Dell'Acqua C., Bò E. D., Benvenuti S. M., and Palomba D. Simone Messerotti Benvenuti, Daniela Palomba, Reduced heart rate variability is associated with vulnerability to depression, *Journal of Affective Disorders Reports*, vol. 1, 2020, 100006, ISSN 2666-9153,https://doi.org/10.1016/j.jadr.2020.100006.

[7] Segerstrom S. C., and Nes L. S. Heart rate variability reflects self-regulatory strength, effort, and fatigue. *Psychol Sci.* 2007 Mar; 18(3):275–81. doi: 10.1111/j.1467-9280.2007.01888.x. PMID: 17444926.

[8] Quintana D. S., Guastella A. J., Outhred T., Hickie I. B., and Kemp A. H. "Heart rate variability is associated with emotion recognition: direct evidence for a relationship between the autonomic nervous system and social cognition." *International journal of psychophysiology: official journal of the International Organization of Psychophysiology* 86 2 (2012): 168–72

[9] Hansen A. L., Johnsen B. H., and Thayer J. F. (2009) Relationship between heart rate variability and cognitive function during threat of shock, *Anxiety, Stress & Coping*, 22:1, 77–89, doi: 10.1080/10615800802272251

[10] Shaffer F. and Ginsberg J. P. "An Overview of Heart Rate Variability Metrics and Norms." *Frontiers in public health* vol. 5 258. 28 Sep. 2017, doi:10.3389/fpubh.2017.00258

[11] Brunoni A. R., Kemp A. H., Dantas E. M. et. al. Benseñor, Heart rate variability is a trait marker of major depressive disorder: evidence from the sertraline vs. electric current therapy to treat depression clinical study, *International Journal of Neuropsychopharmacology*, vol. 16,9 (2013): 1937–1949

[12] Nutt D., Wilson S., and Paterson L. "Sleep disorders as core symptoms of depression." *Dialogues in clinical neuroscience* vol. 10,3 (2008): 329–36. doi:10.31887/DCNS.2008.10.3/dnutt

[13] Triantafillou S., Saeb S., Lattie E. G., Mohr D. C., and Kording K. P. "Relationship Between Sleep Quality and Mood: Ecological Momentary Assessment Study." *JMIR mental health* vol. 6,3 e12613. 27 Mar. 2019, doi:10.2196/12613

[14] Medina A. B., Lechuga D. A., Escandón O. S., and Moctezuma J. V. Update of sleep alterations in depression, *Sleep Science*, vol. 7,3 (2014): 165–169, ISSN 1984-0063

[15] Wichniak A., Wierzbicka A., and Jernajczyk W. "Sleep and antidepressant treatment." *Current pharmaceutical design* 18.36 (2012): 5802–5817.

[16] Baglioni C., Battagliese G., Feige B., et. al. Insomnia as a predictor of depression: a meta-analytic evaluation of longitudinal epidemiological studies. *J Affect Disord*. 2011 Dec; 135(1–3):10–9. doi: 10.1016/j.jad.2011.01.011. Epub 2011 Feb 5. PMID: 21300408.

[17] Wang R., Chen F., Chen Z., et. al. 2014. StudentLife: assessing mental health, academic performance and behavioral trends of college students using smartphones. In Proceedings of the 2014 ACM International Joint Conference on Pervasive and Ubiquitous Computing (UbiComp). Association for Computing Machinery, New York, NY, USA, 3–14. doi: https://doi.org/10.1145/2632048.2632054

[18] Wang R., Wang W., daSilva A., et. al. "Tracking Depression Dynamics in College Students Using Mobile Phone and Wearable Sensing." *Proceedings of the ACM on Interactive, Mobile, Wearable and Ubiquitous Technologies*, vol. 2, no. 1, article 43, Mar. 2018

[19] Palagini L., Baglioni C., Ciapparelli A., Gemignani A., and Riemann D. REM sleep dysregulation in depression: state of the art. *Sleep Med Rev*. 2013 Oct; 17(5):377–90. doi: 10.1016/j.smrv.2012.11.001. Epub 2013 Feb 5. PMID: 23391633.

[20] Giles D. E., Jarrett R. B., Roffwarg H. P., and Rush A. J. Reduced rapid eye movement latency. A predictor of recurrence in depression. *Neuropsychopharmacology*. 1987 Dec; 1(1):33–9. doi: 10.1016/0893-133x(87)90007-8. PMID: 3509065.

[21] Kamen P. W., and Tonkin A. M. Application of the Poincaré plot to heart rate variability: a new measure of functional status in heart failure. *Aust N Z J Med*. 1995 Feb; 25(1):18-26. doi: 10.1111/j.1445-5994.1995.tb00573.x. PMID: 7786239.

[22] Mirescu S. C., and Harden S. (2016). Nonlinear Dynamics methods for assessing heart rate variability in patients with recent myocardial infarction. *Romanian Journal of Biophysics*. 22. 117–124.

[23] Reyes-Zúñiga, M., Castorena-Maldonado, A., Carrillo-Alduenda, J. L., et al. "Anxiety and depression symptoms in patients with sleep-disordered breathing." *The open respiratory medicine journal* vol. 6 (2012): 97–103. doi:10.2174/1874306401206010097

[24] Koldijk S., Sappelli M., Verberne S., Neerincx M. A., and Kraaij, W. 2014. The SWELL Knowledge Work Dataset for Stress and User Modeling Research. In Proceedings of the 16th International Conference on Multimodal Interaction (ICMI '14). Association for Computing Machinery, New York, NY, USA, 291–298. doi:https://doi.org/10.1145/2663204.2663257

[25] Jurewicz, I. "Mental health in young adults and adolescents - supporting general physicians to provide holistic care." *Clinical medicine (London, England)* vol. 15,2 (2015): 151–4. doi:10.7861/clinmedicine.15-2-151

CHAPTER 7

IoT in the Healthcare Sector

A.K. Awasthi, Sanjeev Kumar,
and Arun Kumar Garov

CONTENTS

DOI: 10.1201/9781003227595-7

7.1 INTRODUCTION

Internet of Things (IoT)-enabled services and wearable devices have transformed the lives of patients suffering from chronic diseases and other diseases where patients need intensive care. Mostly, IoT enabled devices are wearable and equipped with activity monitors and measurement devices such as digital scales and digital thermometers. The related data is stored, analysed at the patient's end, and then transmitted through computing clouds to the doctors in the hospital. Patients get feedback from the doctor and make better health and well-being decisions in real-time in a very economical way. The personalized health of the population is greatly benefited in the whole process. Doctors, hospitals, health insurance companies, and artificial intelligence software developers are also providing their services very efficiently. Undoubtedly, the new IoT technology has revolutionized healthcare facilities, but at the same time, service providers are also facing some issues, such as personal sensitive data security, difficulties with regular updates, and global healthcare regulations.

In IoT, things interact and communicate with the users through sensors, microcontrollers, and transceivers. Today, every human being, organization, and institution depends upon the internet for any kind of advancement. IoT-enabled devices use high-quality sensors for smooth and secure communication [1, 2]. The health sector is one of the institutions which is making the best of this novel technique. Since the beginning of the COVID-19 pandemic, the whole world has been focusing on the development of infrastructure in the healthcare sector. IoT has become the backbone of both the service providers and end-users in healthcare facilities. The well-being of an individual has been defined in terms of personalized healthcare, uniquely defined by biological, behavioural, and cultural characteristics. If the right person gets the right help at the right time, then it will be affirmed that we are making the best use of healthcare facilities. But, at present, healthcare services are not very cost effective. IoT ensures the curtailment of highly priced clinical care. The major problem to both the patient and hospital lies with the patient history. IoT-based systems are pervasive, non-invasive, and powerful and thus improve on conventional healthcare systems [3]. In IoT-based healthcare systems, medical information is passed to the cloud through distributed devices and data is collected, stored, and processed at different ends. This data can be made available at any time and at any place with the use of the internet. In this chapter, we will study IoT in healthcare services according to the following topics/section heads (in this chapter IoT and HIoT are used interchangeably):

- IoT in the Healthcare Sector

- Kinds of Technology Used in HIoT

- Importance of IoT in the Healthcare Sector

- Examples of IoT in Healthcare

- The functioning of IoT in Healthcare

- IoT Devices in Healthcare

- Benefits of IoT in Healthcare

- Challenges for IoT in Healthcare

- Services and Applications of an HIoT

- Future Scope of IoT in Healthcare

7.2 IOT IN THE HEALTHCARE SECTOR

The basic topology of IoT in healthcare consists of various components that are coherently interconnected in the healthcare environment, commonly known as HIoT. Processing and dissemination of data with the help of publisher, broker, and subscribers in different stages are shown in

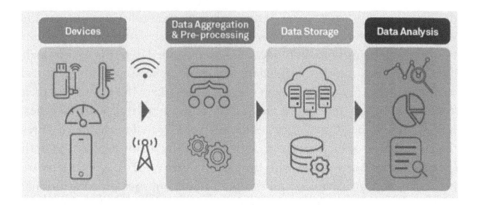

FIGURE 7.1   Stages of IoT solutions.

Figure 7.1. The publisher corresponds to a network of connected sensors which record the patient's important information. The broker receives this information from the publisher continuously and processes the stored data in the cloud. In the final stage, the subscriber continuously monitors the patient's information, which can be accessed with the help of smartphones or laptops. The feedback about the patient's health condition is given by the publisher. This structure is not only the topology that is used everywhere but in the past decade, numerous other IoT architectures have also been proposed for the healthcare environment. The foremost requirement is to follow the medical rules and steps in the diagnosis procedure.

The superlative healthcare facility involves access to a patient's history through high technology and takes remedial measures for further courses of action. Powerful technologies like ultra-fast 5G mobile wireless, artificial intelligence (AI), and big data have revolutionised the usage of IoT in the healthcare sector [4]. Brand essence market research realised that the market of IoT in healthcare will exceed $10 billion at the end of 2024. The remote treatment of patients' health has been revolutionized by the use of 5G technologies in IoT. The usage of IoT is not only helping the patient but also improving the productivity of healthcare workers. Implementation of IoT-based devices has unleashed the potential to keep patients safe and healthy and enriched the capability of physicians to deliver superlative care [5]. It has also increased patient engagement with the health service providers as interactions with doctors have become easier and more efficient. Furthermore, the length of stay and readmissions in hospitals have reduced tremendously with the remote monitoring of a patient's health. The other architecture of IoT in healthcare may consist of four steps, as shown in Figure 7.1.

1. To collect the data, devices like sensors, actuators, monitors, detectors, camera systems etc., are deployed.

2. The data received from the above devices in the same form is converted into digital form for further processing.

3. The digitized and aggregated data is pre-processed and standardized to move further to the cloud.

4. The data received at the final stage is managed and analysed at the appropriate level. For effective decision-making advanced analytics techniques are applied to this data, so that actionable business insights can be generated.

7.3 KINDS OF TECHNOLOGY USED IN HIOT

To enhance the ability and integrate various healthcare applications of IoT in healthcare environments, various ultra-modern technologies have been adopted [6]. These technologies may be classified into three parts, namely, identification technology, communication technology, and location technology (Figure 7.2).

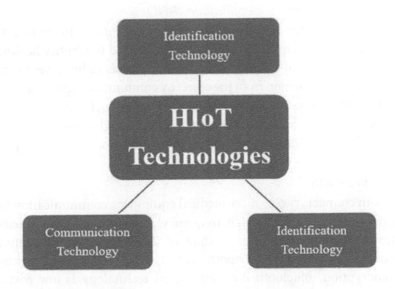

FIGURE 7.2   Kinds of HIoT technology.

7.3.1 Identification Technology

In healthcare IoT/HIoT, a patient's data is accessed at remote locations. This can be carried out by identifying the sensors present in the healthcare system. We can access the patient's data from the authorized node, also known as a sensor. These sensors are present at remote locations. Firstly, a UID (unique identifier) is assigned to each authorized entity and then identification is done so that identified and unambiguous data exchange can be achieved. Universally unique identifiers (UUIDs) and globally developed unique identifiers (GUIDs) are commonly used as well. A UUID, part of a distributed computing environment (DCE), can be operated without the requirement of centralized coordination [7]. For proper functioning of the healthcare system, sensors and actuators are identified separately.

7.3.2 Communication Technology

In an HIoT system, different entities are connected through communication technology. This technology may be short- and medium range. Short-range technology is used to connect objects within a limited range [8]. Medium-range technology supports communication for large distance base stations and body area networks. RFID, Wi-Fi, Zigbee, and Bluetooth are examples of communication technology. Figure 7.3 shows some of the commonly used communication technologies.

7.3.2.1 Radio-Frequency Identification (RFID)

RFID is a short-range communication with a range from 10 cm to 200 m. It is comprised of a tag and a reader. the tag is used to identify healthcare equipment in the HIoT environment. The reader uses radio waves to transmit or receive information from the object by communicating with a tag. The data in the tag are in the form of an electronic product code (EPC). RFID works without any external power. But RFID is a highly insecure protocol and shows compatibility concerns while in a connection with a smartphone.

7.3.2.2 Bluetooth

Bluetooth connects two or more medical equipment communication technologies with UHF (ultra-high frequency) radio waves. The Bluetooth communication has frequency range of 2.4 GHz while communication ranges up to 100 m. Bluetooth protects data through authentication and encryption. Bluetooth communication technology is low cost and energy efficient. Bluetooth connected devices least interferes during data

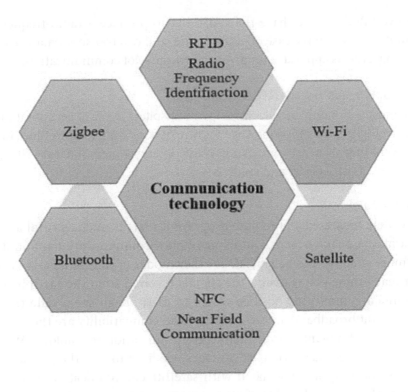

FIGURE 7.3   Communication technologies.

transmission. But, when the healthcare application requires long-range communication, Bluetooth technology fails to serve the purpose.

7.3.2.3 Zigbee

The topology of Zigbee is comprised of end nodes, routers, and a processing centre. Zigbee and Bluetooth have the same frequency range (2.4 GHz) but the former has a higher communication range. This technology is based on mesh network technology. The main features of Zigbee are its highest transmission rate, least power consumption, and high network capacity [9, 10].

7.3.2.4 Near-Field Communication (NFC)

NFC and RFID are similar technologies and use electromagnetic induction for data transmission. There are two operational modes in NFC devices known as the active and passive modes. Only one device generates the radio frequency in the passive mode, while the other device acts as a receiver. Both devices have the capability to produce a radio frequency simultaneously, in the case of active mode, and can transmit data without pairing. NFC is applicable for a very short range of communication.

7.3.2.5 Wireless Fidelity (Wi-Fi)

The most commonly used technology in hospitals is Wi-Fi because it provides a higher transmission range within 70 ft. as compared to Bluetooth. The building of a network using Wi-Fi is very quick and easy. It also ensures robust security and control.

7.3.2.6 Satellite

In widely separated and remote geographical areas, such as rural areas, mountains, peaks, oceans, and so on, satellite communication is found to be more effective and beneficial, as other modes of communication cannot reach these places easily. The signals received from the earth by the satellite are amplified and then resent to earth. High-speed data transfer, instant broadband access, stability, and compatibility are the salient features and advantages of satellite communication technology. While this technology has various advantages, we have to face the problem of power consumption associated with satellite communication, which is very high compared to that of other communication techniques.

7.3.3 Location Technology

The real-time location system (RTLS), also known as location technology, helps us in identifying and tracking the position of an object within the set-up of a healthcare network. RTLS also tracks the treatment processes which are based on the distribution of available resources. One of the most widely used technologies is the global positioning system (GPS). Often, location technology makes use of satellites for tracking purposes. GPS tracks and identifies the objects as long as there exist a clear line of sight between the object and four different satellites. Location technology is employed to detect the position of caregivers, patients, ambulances, healthcare providers, etc. in HIoT set-ups.

7.4 IMPORTANCE OF IOT IN HEALTHCARE

IoT has been becoming an integral part of the healthcare sector. The healthcare industry has adopted this cutting-edge technology most efficiently and quickly. It requires least manual effort and is referred to as an interconnected system of Wi-Fi-embedded sensors and software. To

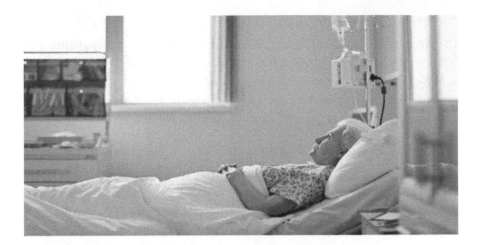

FIGURE 7.4   Patient getting health services through remote monitoring.

monitor the patient's health status 24/7, hundreds of intelligent electronic devices are set up which talk to each other, make decisions, and upload information to a healthcare cloud platform. This enhances the capability and efficiency of healthcare service providers. Remote telemedicine services and medical surgery are major services which are among the top priorities in HIoT.

7.5 EXAMPLES OF IOT IN HEALTHCARE

The fastest growing sector in IoT market is represented by healthcare devices. Therefore, this sector is also known as the Internet of Medical Things (IoMT). There are various examples of IoT in the healthcare industry. The most famous examples are given in the following sections.

7.5.1 Remote Patient Monitoring

The most common application of IoT devices in the healthcare industry is remote patient monitoring. The devices based on IoT can collect health metrics like heart rate, blood pressure, temperature, and many more automatically from patients who are not physically present in a healthcare facility. This eliminates the need for patients to travel to the health service providers.

7.5.2 Glucose Monitoring

IoT devices which monitors glucose levels in patients eliminate the need to keep records manually. Also, these devices send alerts to the patient when glucose levels are not at optimum levels.

7.5.3 Heart-Rate Monitoring

Of late, tiny IoT medical devices monitor the heart rates of patients and freed the patients from moving for cardiological health services. The hearts of these patients are monitored continuously, and necessary dos and don'ts are sent to end-users. Service delivery of these devices has been attributed with 90 percent accuracy.

7.5.4 Hand Hygiene Monitoring

Today, IoT devices remind visitors to sanitize their hands when they enter hospital rooms to meet their kith and kin. These devices also instruct them as to the best way to sanitize the hand and mitigate a particular risk for a particular patient.

7.5.5 Connected Inhalers

The HIoT-connected inhalers help patients who are prone to heart attacks, asthma, COPD, or other problems involving strokes that strike suddenly, with little warning. IoT-connected inhalers monitor the frequency of attacks, as well as collect the data from the environment to help healthcare service providers understand what triggered an attack.

In addition, connected inhalers also alert the patients when they forget inhalers at home and reduce the risk of suffering an attack without their inhaler present.

7.5.6 Ingestible Sensors

To look into the stomach and collect information about the PH levels or any kind of injury which is invasive and causes bleeding, high-quality cameras are used. Cameras with ingestible IoT sensors can travel through the digestive tract to accomplish the same purpose in a less invasive way. Internal bleeding is better seen through these devices but it is not an easy task to put the camera inside the body. Therefore, these devices should be very small so that they can be easily swallowed easily. They must also be able to dissolve or pass through the human body cleanly on their own. Several companies are working on ingestible sensors so that they meet the required standards.

7.5.7 Connected Contact Lenses

For digital interaction, human eyes need smart lenses which can turn human eyes into powerful tools. For example, smart contact lenses provide an opportunity to collect healthcare data in a passive and non-intrusive way. The smart lens wearers can take pictures with the help of their eyes. Whether they're used to improve health outcomes or for other purposes, smart lenses promise to turn human eyes into a powerful tool for digital interactions.

7.5.8 Robotic Surgery

Internet-connected devices or robots when employed inside the human body by surgeons can perform complex procedures which are difficult to manage manually. Robotic surgeries through IoT devices are more promising and less invasive and help the patient to recover faster. These surgeries also reduce the risks associated with incisions. These devices are generally small enough to perform surgeries with the least disruption possible, and they can diagnose complex conditions inside the human body, aiding in making the right decisions.

7.6 THE FUNCTIONING OF IOT IN HEALTHCARE

An IoT set-up can be considered as a system where the sensor interacts with the physical word and sends information through the internet. In IoT healthcare, all the connected devices gather information from patient data and receive input from healthcare service providers. All these devices communicate with each other and inform vital decisions to help and save the life of the patient. Wearable IoT devices make wise and intelligent decisions and call the health service providers in case the condition of patient is critical. In an IoT system all the critical information is sent into the cloud for the doctor to act upon so that timely help, like calling the ambulance, can be arranged. In this way, IoT not only helps the patient but also improves the productivity of health service providers and hospital workflows.

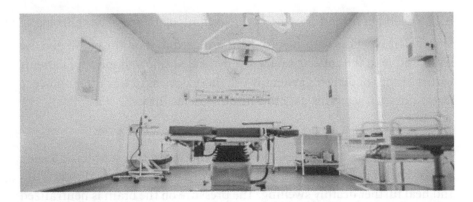

FIGURE 7.5   An IoT set-up in a hospital.

FIGURE 7.6　A smartwatch.

7.7 IOT DEVICES IN HEALTHCARE

A device can be considered an IoT device if it has access to the internet and a radio with a proper TCP/IP address to communicate with other devices through the internet. The simplest example of an IoT device may be a smartphone. The healthcare apps in a smartphone can help us in detecting diseases and improving our health. For example, skin cancer can be detected through the apps using cameras and AI-driven algorithms. Other examples may be yoga, fitness, sleep, and pill management apps. But a dedicated IoT device can be much more than a smartphone.

7.7.1 Smartwatch

Wearable smartwatches come with a sensor and internet connection. For example, the iWatch Series 4 is used in the health industry to monitor heart rate, control diabetes, help in speech treatment, aid in improving posture, and detect seizures.

7.7.2 Insulin Pens and Smart CGM (Continuous Glucose Monitoring)

Smart CGM devices with insulin pens are used to monitor blood glucose levels and send the collected data to a dedicated smartphone app. Patients with a history of diabetes generally use these devices to track their glucose levels and even send this data to a healthcare facilitator.

7.7.3 Brain Swelling Sensors

To track acute brain surgeries, brain surgeons insert tiny sensors and keep a record of brain activities. These sensors are deployed within the cranium and heal further deathly swelling. The pressure on the brain is neutralized automatically in the body without any medical intervention.

7.7.4 Ingestible Sensors

Sometimes a doctor finds it difficult to ensure the medication of patients at all times. Therefore, the doctor prescribes medication with a tiny digestible sensor that, when swallowed by the patient, sends data to a dedicated smartphone app.

7.7.5 Smart Video Pills

When a smart pill travel through a patient's intestinal tract it takes photos. The collected information is sent to a wearable device, which in turn sends the information to a dedicated smartphone app. The gastrointestinal tract and colon can be visualized and tracked through smart pills without any direct interference.

7.8 BENEFITS OF IOT IN HEALTHCARE

The data collected by IoT healthcare devices is highly accurate, and therefore, hospitals, clinic, patients, and other healthcare settings have been greatly benefitted by the use of IoT for better informed decisions. IoT in the healthcare industry has large number of benefits, some of which are mentioned in the following:

- Generally, the facilities and environment in the hospital remain uncomfortable for most of the patients. A recent study in the *Journal of Health Environment Research* shows that privacy, accessibility, security, and comfort are major concerns for some of the families of patients. Study also shows that comfortable and soothing environments can lead to reduced stress and faster recovery. For example:
  - Automated window shades help patients experience the health and mood benefits of a sunny day.
  - Bed sensors monitor the sleeping behaviour of patient and inform the staff about the environment in the room.
  - Wearable devices provide real-time data, trends, and alerts about potential health issues.
- IoMT devices help the physician to make informed medical decisions without physical contact. Elderly patients and patients with Alzheimer's disease or dementia are provided with wearable IoMT devices for their care and safety. Smart belts and smart pills report the health report of patients.

- IoT-based UV light sanitation systems keep the hospital spaces clean and prevent illness.

- IoT-based system set-ups increase the safety of patients, physicians, and staff.

The IOT-based smart hospital can help the future generation with ultra modern facilities.

7.9 CHALLENGES FOR IOT IN HEALTHCARE

IoT in healthcare has a singular potential to improve the health, safety, and quality of life for people almost everywhere. But there are also some challenges that need to be considered when implementing HIoT. Some of the challenges are described in the following:

- **Massive inputs of generated data:** Large numbers of IoT devices are used in a single healthcare facility for sending information from remote locations which ultimately generate large amounts of informative data. The data generated from IoT in healthcare requires storage and also grows much higher, from Terabytes to Petabytes. AI-driven programs and cloud computing helps in organizing this data, but this approach takes a lot of time and is expensive.

- **IoT devices will increase the attack surface:** The healthcare industry has numerous benefits, but also a large number of security concerns. Bad social elements in the form of hackers enter into the medical devices connected to the internet and alter or modify the information for their self-interest. Even an entire hospital can be hacked through infamous ransomware viruses that infect IoT devices. This means hackers could hold patients for ransom by taking control of heart rate, blood pressure readers, and brain mappings put their lives at risk.

- **Existing software infrastructure is obsolete:** Old IT infrastructure in many hospitals are not updated in changing time. Therefore, novel IoT devices are not compatible with the old system. Hence, healthcare facilities need to update their IT processes and use new software. Technologies like SDN, NFV Advanced LTE, or 5G and ultra-fast wireless will help in virtualization.

- **Reliable connectivity:** Network failures can create unacceptable problems in devices that require real-time access to data. To maintain internet connectivity is the most challenging task in wearable mobile devices, which generally travel everywhere with the patient.

- **Cost:** Last but not least is the cost of IoT devices. To benefit all people, use of IoT devices must be economical.

7.10 SERVICES AND APPLICATIONS OF AN HIOT

Recent developments in IoT technology have made it possible for medical service providers to perform remote analysis in real-time. The services and applications of an HIoT have been increasing day by day. Disease remedies, personal care for paediatric and elderly patients, health and fitness management, and supervision of chronic diseases are the topmost applications in HIoT. But the most integral part of any system is its service. The service offered in an IoT environment provides a set of healthcare solutions. IoT services cannot be defined uniquely. Some of these services and applications are shown in Figures 7.7 and 7.8.

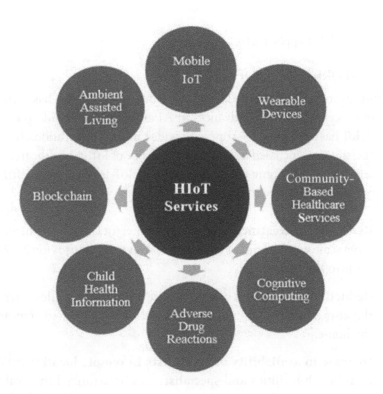

FIGURE 7.7   HIoT services.

FIGURE 7.8   HIoT applications.

7.11 FUTURE SCOPE OF IOT IN HEALTHCARE

The future prospects for IoT are limitless. IoT device automation has been made easy and accessible with the advances in AI and machine and deep learning. Indeed, IoT has broadened its scope with wide ranges of applications, but in the present chapter we focussed on the potential scope of IoT in healthcare.

The following are some of the reasons why IoT is the most useful technology in the healthcare industry:

- **Reduction of treatment error:** Manual errors can be reduced on a large scale with the aid of IoT devices. Devices can provide 24/7 diagnosis to the patient.

- **Reduction of treatment cost:** With the use of IoT, modules can reduce the cost of treatment effectively, as patients are treated remotely in the home.

- **Increase in availability of specialists in remote locations:** World-class health facilities and specialists can be arranged in rural areas in real time.

REFERENCES

[1] Koppar, A. R., & Sridhar, V. (2009). "A workflow solution for electronic health records to improve healthcare delivery efficiency in rural india." In *2009 International Conference on eHealth, Telemedicine, and Social Medicine* (pp. 227–232). IEEE.

[2] Laohakangvalvit, T., & Achalakul, T. (2014, May). "Cloud-based data exchange framework for healthcare services." In *2014 11th International Joint Conference on Computer Science and Software Engineering (JCSSE)* (pp. 242–247). IEEE.

[3] Ambarkar, S. S., & Shekokar, N. (2020). *Toward smart and secure IoT based healthcare system; Internet of things, smart computing and technology: A Roadmap Ahead* (pp. 266, 283–303). *Springer, Cham*.

[4] Ko, Jeong Gil, Chenyang, Lu, Srivastava, B. M., & Stankovic, J. (2010). "Wireless Sensor Networks for Healthcare." *Proceedings of the IEEE*, 98(11), 1947–1960.

[5] Matta, P., & Pant, B. (2019). "Internet of things: Genesis, challenges and applications." *Journal of Engineering Science and Technology*, 14(3), 1717–1750.

[6] Tan, J., & Simon, G. M. K. (2014). "A survey of technologies in internet of things." In *IEEE International Conference on Distributed Computing in Sensor Systems*.

[7] Coetzee, L., & Eksteen, J. (2011, May). "The Internet of Things-promise for the future? An introduction." In *2011 IST-Africa Conference Proceedings* (pp. 1–9). IEEE.

[8] Mufti, T., Sami, N., Sohail, S. S., & Neha. (2020). "Future Internet of Things (IOT) from cloud perspective: Aspects, applications and challenges." In *Internet of Things (IoT)* (pp. 515–532). *Springer, Cham*.

[9] Kulkarni, A., & Sathe, S. (2014). "Healthcare applications of the Internet of Things: A review." *International Journal of Computer Science and Information Technologies*, 5(5), 6229–6232.

[10] Bhat, S., Bhat, O., & Gokhale, P. (2018). "Applications of IoT and IoT: Vision 2020." *International Advanced Research Journal in Science, Engineering and Technology*, 5(1), 41–44. doi:10.17148/IARJSET.2018.516.

CHAPTER 8

The Role of IoT in Sustainable Healthcare

Ashish Mulajkar, Sanjeet K Sinha, Vinod Bharat, Arundoy Lenka, and Govind Singh Patel

CONTENTS

8.1 INTRODUCTION TO IOT

The rise of Internet of Things has potentially improved lifesaving applications in the healthcare industry. By collecting data from bedside devices, patient information and real-time data diagnosis can be improved. By 2090, around 70 percent of healthcare organizations shall implement IoT technology [1]. IoT seems to be an initiative to create efficiency, potential to retrieve data loss, and reduce mistakes in diagnosis. Over 50 percent of devices on healthcare networks in the next two years

DOI: 10.1201/9781003227595-8

will be IoT devices [1]. From hand-held devices and health records to medical equipment, the industry is embracing the world of connected things. Diagnosis of any critical disease using IoT is analysed using standard optimization algorithms with profound efficiency. The expert can give remote diagnosis and track medical assets without manually visiting each patient. Using appropriate sensors, Wi-Fi, and modules, the ability to locate the right department in a hospital becomes easy for both caregivers and patients. This chapter provides a detailed study of various advanced IoT-based sustainable healthcare technologies like blockchain, fog computing, big data, and machine learning techniques which can reduce stress on the current healthcare sector. This chapter also briefly describes integrated actions in Aruba.

As the population and pollution in the world increases, people mostly suffer from common diseases like asthma, cancer, AIDS, psoriasis, and COPD [1], and the number is growing with a linear speed. MSDs (musculoskeletal disorders) are the defects which causes issues in muscular movements of the body parts [2]. To avoid the growth of MSD, an early detection technique called Rodgers Muscle Fatigue Analysis was developed. Air pollution due to heavy industrialization in urban areas has caused many premature diseases and deaths in recent years. The death count is up to five million worldwide per year, making it fifth amongst all disorders [3]. Many healthcare models are shifting towards patient-centric methods, common among which is ICT (information and communication technologies) along with IoT platforms [4]. Air Quality Index (AQI)-detecting stations should be installed in urban areas which in turn help to build smart cities. WSNs (Wireless Sensor Networks) and some advanced sensors are used in conventional air quality measurements to avoid high costs and power consumption. Table 8.1 lists AQI specifications.

TABLE 8.1 AQI Index Specifications

AQI Limits	Colour	Illustration
0–33	Blue	Normal
34–66	Green	Normal
67–99	Yellow	Moderate. Respiration problems to sensitive peoples
100–149	Orange	Outdoor activities to be rescheduled
150–200	Purple	Sensitive groups to get some additional disorders
200 +	Red	Warnings: Serious health issues may occur regarding respiratory disorders

8.2 LITERATURE SURVEY AND RELATED WORK

Laizhong Cui et al. [5] have concentrated their study on various applications available in machine learning (ML), which can be used in IoT. They have considered various parameters of IoT applications such as device identification, security of devices, and edge computing.

Jithin Jagannath et al. [6] have focused their study on considering the ad hoc network view in IoT. Basically Jithin Jagannath et al. explain all the basic techniques of ML which are useful the study and concentrated on the various levels of the networking such as physical, data link, and network.

David Gil et al. [7] explore the study of big data which is quite useful in the field of IoT. They also focus on various ML and deep learning (DL) techniques, such as the neural network, which are perfectly suited for big data.

Ranesh Kumar Naha et al. [8] provide insight about the most recent technology, namely, fog computing. Their study focuses on the very basic details of fog computing and elaborates on its various trends, outlining the differences between fog and cloud computing. Mainly the study provides insights about the various components of the architecture of fog computing.

Dr. D. Sivaganesan [9] has introduced the concept of blockchain into IoT platforms for secure transmission of data from source to destination in the field of medical science, also arguing that transparency of data can be achieved using different sensors connect to the cloud.

8.3 IOT IN SMART HEALTHCARE SYSTEMS

Transforming the conventional healthcare sector into an IoT platform can automate the tasks performed by humans. IoT also enriches new techniques like AAL (ambient assisted living), e-health (electronic health), and m-health (mobile health) which helps to supervise such patients remotely and maintain records via the cloud [10]. Figure 8.1 illustrates general elements used in IoT healthcare systems.

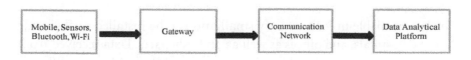

FIGURE 8.1 General elements used in IoT healthcare.

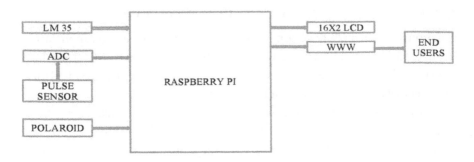

FIGURE 8.2 Smart health monitoring system based on Raspberry Pi.

With the increase of broad areas of IoT, connecting devices and wearables to the cloud has become easy. As a result, analysis of data from different resources is possible. In the context of smart cities, the healthcare sector using IoT methods will be used in almost all sections. This gives rise to an advanced technique called smart health (s-health) based on IoT [11]. Figure 8.2 illustrates the smart health monitoring system based on Raspberry Pi.

8.4 COMMUNICATION TECHNIQUES IN IOT HEALTHCARE

Healthcare applications using IoT experience some distortions and challenges. Precise and verified data received from the sensors need to be thoroughly verified. As per ITU (International Telecommunication Unit) standards, the architecture network in design of IoT includes five different layers [12].

1. Sensing

2. Access

3. Network

4. Middleware

5. Application

Sensors available in IoTs are very small and can be installed easily inside houses, hospitals, and offices as well as on body parts. Data received from the sensors can be analysed in a broader area while considering different parameters. The use of non-static architecture design is recommended to

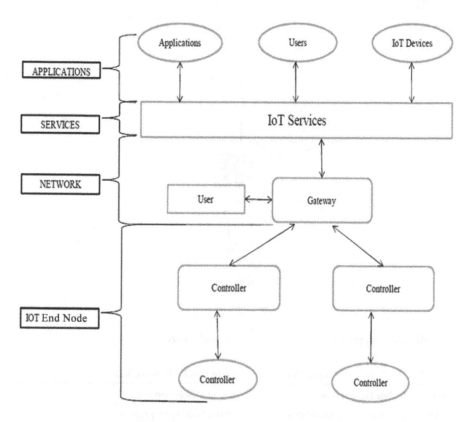

FIGURE 8.3   Architecture for IoT healthcare applications

analyse data received from number of sensors in IoT-connected devices [13]. Compared to open system interconnects (OSI) reference model in communication networks, these layers exist in IoT design models, as shown in Figure 8.3.

8.5 ADVANCED IOT TECHNOLOGIES

8.5.1 Use of Blockchain in IoT

The implementation of blockchain makes IoT a safer platform for securing the sensitive data. Mostly blockchain techniques are used in financial sectors to provide a secure platform. It is an emerging technology extensively used in IIoT (Industrial Internet of Things) and Industry 4.0 [14]. Blockchain is also extensively used in crypto currency areas like Bitcoin and Ethereum. To maintain the record of each transaction a proper balance sheet is used. Other applications of blockchain include the educational

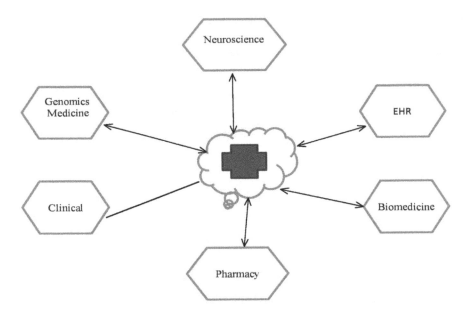

FIGURE 8.4   Application of blockchain in healthcare.

TABLE 8.2   Related Literature Surveys regarding Block-chain for IoT

Year	Author Description	Area of Discussion
2018	Valentina Lenarduzzi et al. [23]	Analysis of Agile projects
2019	Tejasvi Alladi et al. [14]	Applications in terms of IIoT and Industry 4.0
2019	Ali Alammary et al. [24]	Educational applications of Block-chain
2020	M.P. McBee et al. [25]	Medical Imaging principle and application
2021	R. Banach et al.	Applications of Block-chain in Crypto coins

sector, manufacturing sector, and the supply chain. To reduce the faults in critical patient records in healthcare, blockchain provides vital solutions [15]. Figure 8.4 shows various applications of blockchain in the healthcare sector. Table 8.2 indicates related literature surveys for blockchain.

8.5.2 Fog Computing in IoT

The unique features of fog computing in healthcare include lower latency with respect to cloud computing, real-time applications, and improved response times [16]. A new advanced concept called "identity and access management (IAM)" is introduced to maintain the privacy and security of data and to properly analyse the data. All the sensors located in different layers of the fog can be connected to bridge the gap between the smart

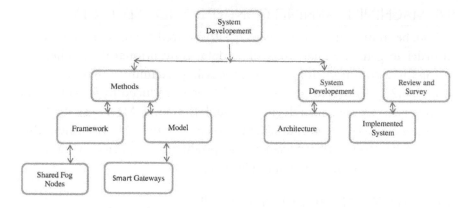

FIGURE 8.5   Healthcaretaxonomy in fog computing.

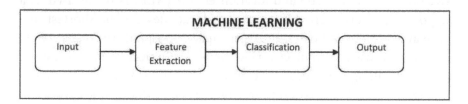

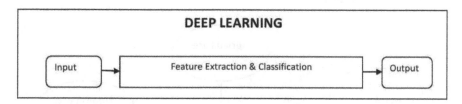

FIGURE 8.6   Machine learning vs. Deep Learning

e-healthcare system and the client [17]. The parameters of healthcare taxonomy in fog computing are shown in Figure 8.5.

8.5.3 Big Data in IoT

It has become necessary to combine prevailing technologies with new ones like big data and ML to overcome threats to IoT [18]. Data handling capacity of a system improves when large amounts of data received from various ends are combined. DL is a subsidiary of ML, which has three major components: supervised, semi-supervised and un-supervised. It includes different layers of ANN (artificial neural networks). Each layer consists of neurons like those of human brains, which produce non linear outputs [19]. Figure 8.6 shows the difference between ML and DL.

8.6 MACHINE LEARNING CONCEPT IN IOT SECURITY

IoT can be communicated by means of either wired or wireless technology. In order to generate huge amount of data, conventional data collection process like common storage and processing techniques may not work properly [20]. To overcome this issue, ML is introduced to provide extra intelligence to IoT-enabled devices through the cloud. ML is a smart technique introduced to communicate between humans and machines. This technique improves the efficiency and production capacity in industry. The role of ML in IoT is shown in Figure 8.7.

8.7 INTEGRATED ACTIONS IN ARUBA

To connect the preferred devices among themselves in a building, the ARUBA network analytic and location engine (ALE) is being used as a supportive platform [21]. ALE is used to locate devices the shortest distance away for the end users. In larger capacity hospitals in Aruba, Wi-Fi technology is used to connect various devices. Physicians can track wireless phones along with variables by means of IoT in Aruba [22]. For setting up extended Wi-Fi coverage in larger areas, the technique adopted in ARUBA is widely used.

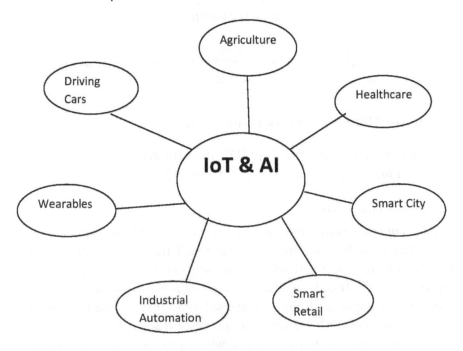

FIGURE 8.7   Role of machine learning in IoT.

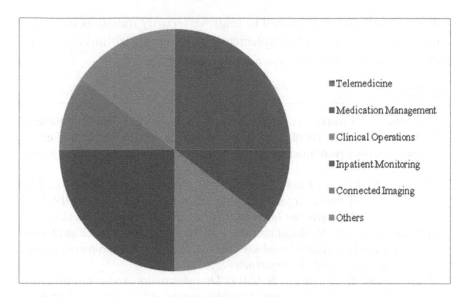

FIGURE 8.8   Healthcare market size analysis using IoT. (Source: IoT in health-care market size report 2019–2025. https://www.grandviewresearch.com/industry-analysis/internet-of-things-iot-healthcare-market.)

8.8 RESULTS

The IoT healthcare market globally is valued at $147 billion and is about to increase by about 19.9 percent in coming years [23]. The key elements for improving the growth of hospitals, pharmacy industry, wearables adoption of block chain, data science, and fog computing are widely used [24]. IoT solutions most adopted by medical practitioners have been Doctor on Demand, Teladoc, and iCliniq [25]. The Pie chart predicts the analysis of various medical components and their its market share using IoT (Figure 8.8).

8.9 CONCLUSION

The infrastructure of smart cities can be improved only if the urban health-care facilities are developed. Using IoT and allied techniques like block-chain, fog computing, ML, and AI, the current scenario of the healthcare sector can be improved. Using s-health) and m-health techniques with IoT, physicians can easily deploy the data of patients from different sensors. This chapter also briefly discussed integrated actions in Aruba and their implementation.

Future work on IoT is needed to improve security measures in healthcare for the safety of data by implementing various algorithms along with platforms like ML, DL, big data, and fog computing.

REFERENCES

[1] Casino, F., Patsakis, C., Batista, E., Postolache, O., Martínez-Ballesté, A., & Solanas, A. (2018, September). Smart healthcare in the IoT era: A context-aware recommendation example. In *2018 International Symposium in Sensing and Instrumentation in IoT Era (ISSI)* (pp. 1–4). IEEE.

[2] Low, J. X., Wei, Y., Chow, J., & Ali, I. F. (2019, July). ActSen-AI-enabled real-time IoT-based ergonomic risk assessment system. In *2019 IEEE International Congress on Internet of Things (ICIOT)* (pp. 76–78). IEEE.

[3] Senthilkumar, R., Venkatakrishnan, P., & Balaji, N. (2020). Intelligent based novel embedded system based IoT enabled air pollution monitoring system. *Microprocessors and Microsystems, 77*, 103172.

[4] Majeed, M. T., & Khan, F. N. (2019). Do information and communication technologies (ICTs) contribute to health outcomes? An empirical analysis. *Quality & quantity, 53*(1).

[5] Cui, L., Yang, S., Chen, F., Ming, Z., Lu, N., & Qin, J. (2018). A survey on application of machine learning for Internet of Things. *International Journal of Machine Learning and Cybernetics, 9*(8), 1399–1417.

[6] Jagannath, J., Polosky, N., Jagannath, A., Restuccia, F., & Melodia, T. (2019). Machine learning for wireless communications in the Internet of Things: A comprehensive survey. *Ad Hoc Networks, 93*, 101913.

[7] Gil, D., Johnsson, M., Mora, H., & Szymański, J. (2019). Review of the complexity of managing big data of the Internet of Things. *Complexity, 2019.*

[8] Naha, R. K., Garg, S., Georgakopoulos, D., Jayaraman, P. P., Gao, L., Xiang, Y., & Ranjan, R. (2018). Fog computing: Survey of trends, architectures, requirements, and research directions. *IEEE Access, 6*, 47980–48009.

[9] Sivaganesan, D. (2019). Block chain enabled internet of things. *Journal of Information Technology, 1*(1), 1–8.

[10] Rahmani, A. M., Gia, T. N., Negash, B., Anzanpour, A., Azimi, I., Jiang, M., & Liljeberg, P. (2018). Exploiting smart e-health gateways at the edge of healthcare Internet-of-Things: A fog computing approach. *Future Generation Computer Systems, 78*, 641–658.

[11] Rahaman, A., Islam, M., Islam, M., Sadi, M., & Nooruddin, S. (2019). Developing IoT based smart health monitoring systems: A review. *International Information and Engineering Technology Association, 33*, 435–440.

[12] Vijayakumar, K., & Bhuvaneswari, V. (2020, February). A Ubiquitous first look of IoT framework for healthcare applications. In *2020 International Conference on Emerging Trends in Information Technology and Engineering (ic-ETITE)* (pp. 1–7). IEEE.

[13] Wu, F., Wu, T., & Yuce, M. R. (2019, April). Design and implementation of a wearable sensor network system for IoT-connected safety and health applications. In*2019 IEEE 5th World Forum on Internet of Things (WF-IoT)* (pp. 87–90). IEEE.

[14] Alladi, T., Chamola, V., Parizi, R. M., & Choo, K. K. R. (2019). Blockchain applications for industry 4.0 and industrial IoT: A review. *IEEE Access*, 7, 176935–176951.

[15] Esposito, C., De Santis, A., Tortora, G., Chang, H., & Choo, K. K. R. (2018). Blockchain: A panacea for healthcare cloud-based data security and privacy? *IEEE Cloud Computing*, 5(1), 31–37.

[16] Mutlag, A. A., Abd Ghani, M. K., Arunkumar, N. A., Mohammed, M. A., & Mohd, O. (2019). Enabling technologies for fog computing in healthcare IoT systems. *Future Generation Computer Systems*, 90, 62–78.

[17] Rajagopalan, A., Jagga, M., Kumari, A., & Ali, S. T. (2017, February). A DDoS prevention scheme for session resumption SEA architecture in healthcare IoT. In *2017 3rd International Conference on Computational Intelligence & Communication Technology (CICT)* (pp. 1–5). IEEE.

[18] Amanullah, M. A., Habeeb, R. A. A., Nasaruddin, F. H., Gani, A., Ahmed, E., Nainar, A. S. M., & Imran, M. (2020). Deep learning and big data technologies for IoT security. *Computer Communications*, 151, 495–517.

[19] Zhou, Y., Han, M., Liu, L., He, J. S., & Wang, Y. (2018, April). Deep learning approach for cyberattack detection. In *IEEE INFOCOM 2018-IEEE Conference on Computer Communications Workshops (INFOCOM WKSHPS)* (pp. 262–267). IEEE.

[20] Hussain, F., Hussain, R., Hassan, S. A., & Hossain, E. (2020). Machine learning in IoT security: Current solutions and future challenges. *IEEE Communications Surveys & Tutorials*, 22(3), 1686–1721.

[21] Sattarian, M., Rezazadeh, J., Farahbakhsh, R., & Bagheri, A. (2019). Indoor navigation systems based on data mining techniques in internet of things: A survey. *Wireless Networks*, 25(3), 1385–1402.

[22] Sedes, Florence (Ed.). (2018). *How Information Systems Can Help in Alarm/ Alert Detect.*, [Elsevier.

[23] Lenarduzzi, V., Lunesu, M. I., Marchesi, M., & Tonelli, R. (2018, May). Blockchain applications for Agile methodologies. In *Proceedings of the 19th International Conference on Agile Software Development: Companion* (pp. 1–3).

[24] Alammary, A., Alhazmi, S., Almasri, M., & Gillani, S. (2019). Blockchain-based applications in education: A systematic review. *Applied Sciences*, 9(12), 2400.

[25] McBee, M. P., & Wilcox, C. (2020). Blockchain technology: Principles and applications in medical imaging. *Journal of Digital Imaging*, 33(3), 726–734.

CHAPTER 9

EMG-Based Robot Control Human Interfaces for Hospitals

Govind Singh Patel, Dhiraj Gupta, Baibaswata Mohapatra, and Sunil Kumar Chaudhary

CONTENTS

DOI: 10.1201/9781003227595-9

9.1 INTRODUCTION

This chapter demonstrates and presents robotic limb movement that generates electrical signals to stimulate and rehabilitate the EMG signals. The behaviour of robotic arm for metrical task has been observed. This concept can be used in many applications, like tele-operating a truss structure, air flight control, space robotics missions, and other rehabilitation aids.

Electromyogram signals are generated by flexible and robust communication techniques with appropriate commands for corresponding movement. Such types of humanoid robots will be used to assist for maintaining the environment.

There are three aspects to finding the target. The first aspect deals with how the embedded system [1–3] chronology (brain of robot) can explore its function. The second aspect is how to move the corresponding arm with their actual quantity. Finally, the third aspect is how to realize the program according to movement. In the next section, a modelling circuit diagram of the proposed technique is explained.

9.2 MODELLING CIRCUIT DIAGRAM

In this section, a modelling circuit of the proposed technique is shown (Figure 9.1). It has many blocks, but EMG and Robotics are two main blocks. Let's discuss each of them in detail in the next subsections.

9.2.1 Subject

The subject is connected to the EMG unit which provides a human being's signal to the EMG, which is attached with an EMG sensor, and raw EMG signals are acquired from the subject [4]. Sensors having two surface electrodes with small distance between two electrodes are placed on the muscle to pick up the raw signals with their corresponding actions. These raw signals are very weak so these signals need amplification for the further process [5–10].

9.2.2 Electromyogram Sensors

A sensor is a device that produce electrical signals from a physical quantity. There are many types of sensors, but here an EMG sensor is used with either data link or data log to collect information from muscle activity, as shown in Figure 9.2. The model name of this sensor is SX230.

This model needs amplification and it is manufactured by Biometrics. It is user-friendly and generates good quality signals for further process. Input impedance of the sensor is greater than 10,000,000 M Ohms. There is no need to prepare the skin with cream to record good signal from muscles with static and dynamic applications in medical science.

Measurement of EMG signal over human body gets very simple using double-sided tapes. This type of sensors needs one reference cable for the ground connectivity.

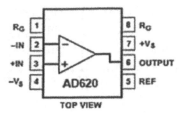

FIGURE 9.1   Instrumentation amplifier.

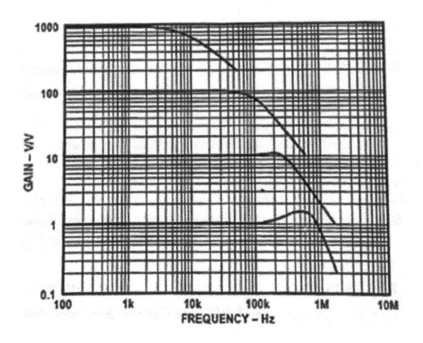

FIGURE 9.2   Gain versus frequency.

9.2.2.1 Electromyogram Sensors

The advantages of EMG sensors are as follows:

1. Both electrodes are used to detect low-level signals in any noisy surrounding environment.

2. Due to high input impedance (>10,000,00 M Ohms) to reduce mismatches of skin contact resistance when difference of pickup inputs have been measured.

3. Other commercially available systems are also using this sensor to minimize noise in static and dynamic applications.

4. The biometrics sensor includes a third-order filter.

5. DC offset is removed by a high-pass filter which is generated by membrane potentials.

6. The unwanted frequencies above 360 Hz have been removed by low-pass filters.

7. This sensor model has and eighth-order elliptic filter.

8. It has a low-noise amplifier with a CMMR ratio of 115 dB.

9.2.3 Electromyogram Unit

This unit consists of an instrumentation amplifier and filter section. Electromyogram unit generates raw signal, which is used by EMG signal [11–13]. EMG unit will process these signals to the controller for realization in a convenient form.

9.2.4 Instrumentation Amplifier

This is type of amplifier is used to amplify weak raw signals into the controller for further process. It consists of three operational amplifiers, is shown in Figure 9.1.

Two amplifiers are working in non-inverting mode with their inverting terminals not grounded. Inverting terminals are connected with a feedback loop. The third amplifier will work as differential amplifier. It has a high-input impedance and high CMMR.

9.2.4.1 Instrumentation Amplifier

9.2.4.1.1 Advantages of Instrumentation Amplifier

1. Common mode rejection ratio is high

2. Bandwidth must be greater than 110k Hz

3. Gain should be 1000 or more

Gain versus frequency and CMR versus frequency are shown in Figures 9.2 and 9.3. Let us consider that EMG has a low amplitude, and so total gain should be around 2000. There is the possibility of adding a second-stage operational amplifier (gain 20), which would garner a gain of around 100. The following are steps to use to configure the amplifier:

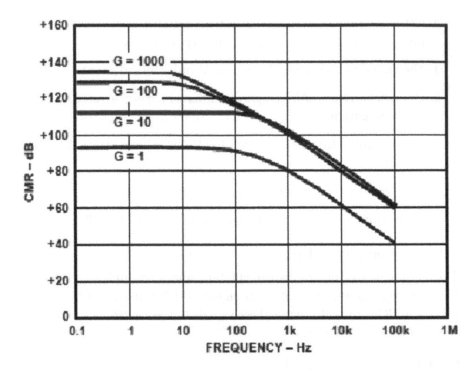

FIGURE 9.3   CMR versus frequency.

Step 1. Build and test the stage of AD620

Step 2. Build operational amplifier stage

Step 3. Join stages

Step 4. Join alternative current coupling

Step 5. Attach electrodes from the amplifier to a human muscle

Step 6. Detect electromagnetic signals from muscle

Step 7. Add notch filter and diode detector

Step 8. Add LPF for smoothing EMG signal

The gain of the amplifier is shown in Figure 9.2.

9.2.5 Filter Section for Processing Unit

A notch filter is used to filter inherent noise at 0 Hz signal. The band pass filter is used to filter out frequencies between 12 Hz and 5 KHz from the

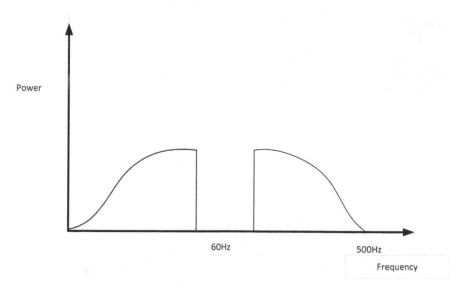

FIGURE 9.4   Signal processing unit.

original signal. It is known that the tibialis anterior muscle gives large and clear signals to the output. But any biceps of the arm can be used for the same [14–15].

To reduce noise to the lowest level, attach electrodes close together. Special type (in pairing form) EMG electrodes can also be used. Other electrodes (common, ground) can be connected to an area like a knee cap or shin bone where there is no muscle. The various extracted features are shown in Figure 9.4.

9.2.6 Bandstop and Bandpass Filters

The frequency content of an EMG signal is in the range of 240 to 500 Hz, while for an EKG the signal is around 40 Hz. This means that frequency content of EMG is larger than that of EKG. In that case, a filter will adjust the upward frequency as required. By this mechanism, AC electrical noise will not be eliminated. To eliminate this noise, a bandstop filter or notch filter is used in a narrow frequency range.

Frequency ranges of the notch filter depending upon the order of filter and cut-off frequency are shown in Figures 9.5 and 9.6. For designing this type of filter, passive components with twin circuits can be used.

Practically, it is very difficult to get accurate values for the circuit to tune the circuit. The values of the resistors should be adjusted simultaneously.

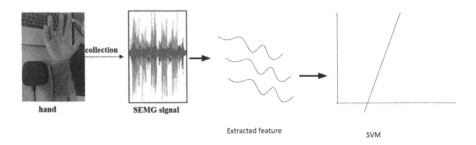

hand   SEMG signal   Extracted feature   SVM

FIGURE 9.5   Bandpass filter.

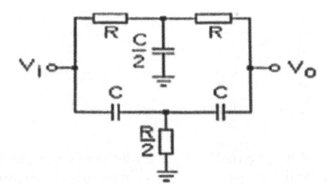

FIGURE 9.6   Twin circuit.

One preset is used to tune the frequency of the notch filter (alternative filter), and a second preset is used to tune its depth for maximum attenuation.

9.2.7 Making the EMG Signal Resemble the Muscle Force

Initially force and timing of muscle contraction can be measured by the instrument. For understanding force and EMG, difference in the variables can be determined for further process.

A muscle used to generate force with contractile property is shown in Figure 9.7. This force has a frequency up to 3–6 Hz.

The conversion of a raw EMG signal into resembling signals requires two steps:

1. Detection

2. Smoothing

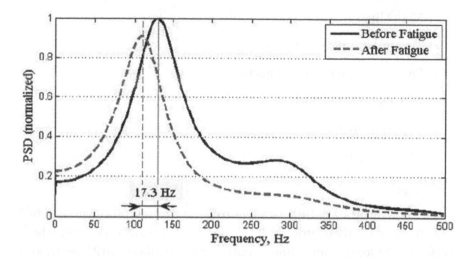

FIGURE 9.7   Graph of gastrocnemius

FIGURE 9.8   IC and block diagram of 8051 microcontroller.

9.3 PROCESSING UNIT

A microcontroller is a highly integrated device that includes a CPU, memory, timers, interrupts, input/output port, etc [1]. A block diagram and IC diagram are shown in Figure 9.8.

The basic features of the microcontroller are:

- 4 K bytes of flash memory

- 128 bytse of RAM

- 32 I/O lines

- Two 16-bit timers/counters

- Five vector interrupt architecture

- Full duplex serial port

- "On chip" oscillator

- Clock circuitry

9.3.1 CPU

The CPU is the heart of the microcontroller. It controls all the peripherals and memory for their process. It is also called a main processor, which performs the logic, controlling, input/output operations, arithmetic, and other operations by the specified instructions.

9.3.2 RAM

RAM is random access memory. It is a volatile memory. It is used to store information temporary. It is a hardware that stores operating system and other application software. It also helps to access data quickly. It is being used for booting of the system. RAM increases processing speed. Higher capacity RAM is required to run software as well as video games.

There are two types of RAM:

1. Static

2. Dynamic

9.3.3 ROM

ROM is read only memory. It is a non-volatile memory. It is used to store data permanently. Many types of ROM are available on the market.

Flash Memory is a type of ROM memory. It is used to store data permanently. Complete data can be erased at a time and rewritten over this memory. Generally this type of memory is used in embedded systems or IoT-based applications where less memory and a smaller budget is required. Programmer is a device which is used to complete the process.

9.3.4 PROM

PROM is programmable ROM memory. It is also non-volatile memory. Many embedded system applications can be programmed/designed using this type of memory. It consists of a decoder and OR gates.

9.3.5 EPROM

EPROM is electrical programmable erasable ROM memory. It is also non-volatile memory. It is also used to design embedded based applications. In this type of memory, we cannot erase data at any time. Using this type of memory, data can be erased track/sector wise with the help of a programmer.

9.3.6 Timers

A timer is a device that can be used either as a timer or counter. Timers are used to generate delays in any application. And as a counter, timers are used to count the number of clock pulses arriving at clock input.

The microcontroller 8051 has two 16-bit times and TMOD registers:

- Timer 0: TH0—Higher bits; TL0—Lower bits

- Timer 1: TH1—Higher bits; TL1—Lower bits

TMOD (Timer mode) register is an eight-bit register that can be used to configure different modes with timers as per the requirement of the applications in the field of embedded systems or other IoT-based applications. It also determines how to start/stop timers for their respective applications.

9.3.7 SFRs (Special Function Registers)

SFRs are those registers used to store data for the further process. These registers occupy 128 bytes of RAM of the microcontroller. Their locations are fixed in RAM; they are also accessed by their memory address.

9.3.8 General Purpose Registers

These registers are used to store or shift the data from one place/memory to another place/memory for performing arithmetic, logical, shift, and input/output operations.

Example: A, B, R0, R1, . . . PC, SP . . . . . .

9.3.9 Register Banks

The microcontroller 8051 has four banks and each bank has eight registers: R0, R1, R2, R3, R4, R5, R6, R7. Data of these registers can be accessed by name and by addresses. They take up 32 bytes of memory to store the data in RAM.

9.3.10 Serial Input/Output

Microcontroller 8051 has serial input and output ports to communicate data serially. A serial input port is used to send data serially to the other devices. A serial output port is used to receive data serially from other external devices to the microcontroller. For transmitting or receiving the data, a serial buffer eight-bit register is used to communicate the information from one place to another. Data can be transmitted/received only by an eight-bit SBUF register not by other registers.

9.3.11 I/O Port

Microcontroller 8051 has 32 I/O ports along with four eight-bit ports. These ports can be used either for input or output to access data and to connect various peripherals for further communication.

9.3.12 Control Unit

A control unit is used to control all the functions performed by the microcontroller with various devices and peripherals. Data bus accesses data with the permission of the CPU and control unit. The frequency of the oscillator is used to synchronize the devices and peripherals for their respective applications.

9.3.13 Interrupts

Interrupts are request signals that can be used to utilize free time of the process for the interrupt applications.

There are two types of interrupts: software interrupts and hardware interrupts.

9.3.13.1 Software Interrupts

Software interrupts are those which can be controlled by software. Examples of interrupts are RST 7.5, RST 6.5, RST 5.5, etc.

9.3.13.2 Hardware Interrupts

Hardware interrupts are those which can be controlled by hardware, such as INTR, etc.

Classifications of interrupts are based on enable or disable properties:

9.3.13.3 Maskable Interrupts

Maskable interrupts are those which can be enabled/disabled by software or hardware, such as RST 7.5, RST 6.5, RST 5.5, etc.

Non-Maskable interrupts are those which cannot be enabled/disabled by software or hardware, such as TRAP.

9.3.14 Pin Description of 8051 Microcontroller

Pins nos. 1–8: These pins can be configured as an input or output individually.

9.3.14.1 Pin No. 9 (RS, Reset)

If it has a logic of 1 then the microcontroller stops processing and the content of all the registers is erased. And if it has a logic of 0 then the microcontroller starts execution from the beginning. It means that microcontroller is reset if positive voltage pulse is applied.

9.3.14.2 Pins Nos. 10–17 [Port 1]

These pins can be used as an input or output pins individually. Each pin has a few functions to perform alternate tasks.

9.3.14.3 Pin No. 10 [RXD]

This is used to communicate data serially. It can be used as serial synchronous communication output or serial asynchronous communication input.

9.3.14.4 Pin No. 11 [TXD]

This is used to communicate data serially. It can be used as serial synchronous communication clock output or serial asynchronous communication output.

9.3.14.5 Pin No. 12 [INT0]

This works as an interrupt 0 input.

9.3.14.6 Pin No. 13 [INT1]

This works as an interrupt 1 input.

9.3.14.7 Pin No. 14 [T0]

This works as a counter 0 clock input.

9.3.14.8 Pin No. 15 [T1]

This works as a counter 1 clock input.

9.3.14.9 Pin No. 16 [WR]

This signal is used to write data to external additional RAM.

9.3.14.10 Pin No. 17 [RD]

This signal is used to read data from external additional RAM.

9.3.14.11 Pins Nos. 18, 19 [X2,X1]

These pins are used to access internal oscillator inputs and outputs. These pins are connected to quartz crystals to provide operating frequency for the microcontroller. Other types of microcontrollers use over 50 Hz crystal frequencies for the operations.

9.3.14.12 Pin No. 20 [GND]

This pin uses as ground for the respective applications.

9.3.14.13 Pins Nos. 21–28 [Port 2]

These pins can be used as input or output pins individually. Each pin also has a few functions to perform alternate tasks.

9.3.14.14 Pin No. 29 [PSEN]

If this has a logic of 0 then external ROM is used to store the program. It is also called a program store enabled pin.

9.3.14.15 Pin Nos. 30 [ALE]

The address latch enable (ALE) pin is used to access data and address bus at particular clock cycles at the specified time. Single bus is used to share both the information address and data simultaneously. In first one and half cycle, address is being accessed, and after that data is being accessed for the further process.

9.3.14.16 Pin No. 31 [EA]

The external access (EA) pin is used to access devices with the coordination of 4288 IC. It configures external peripherals with the processor. Multiple microcontrollers may also be configured using this IC.

9.3.14.17 Pins Nos. 32–39 [Port 0] AD0–AD7

These pins can be used as input or output pins individually. These pins are also called multiplexed by which we can access data as well as address.

9.3.14.18 Pin No. 40 [VCC]

This pin is used to supply 5 V to the microcontroller.

9.4 DRIVER IC

The driver IC is used to interface between the microcontroller and robotic arm. The robotic arm consists of DC motors to control the movement of arms corresponding to the action performed [6–8] by the arm. Many driver ICs are available on the market to control the arm, but in this work the H-bridge driver integrated circuit is used for switching purposes. The name of this IC is L293D. It has 16 pins and four channels with high voltage and high current based on diode transistor logic (DTL) technology. These technologies are generally used to control solenoid relays, switching power transistors, DC motors, stepper motors, etc. Switching applications up to 5 KHz can be controlled by this device, as shown in Figure 9.9.

The driver IC has four channel drivers with a delivering capacity of around 595 mA per channel. It can control motion of the motors in the direction of stop/forward/reverse, etc. These motions or directions can be controlled by microcontrollers such as PIC, AVR, ARM, etc.

The basic features of the L293D IC (Figure 9.10) are:

- Per driver output current capacity is 600 mA

- 1.2 amps per pulse current

- Voltage range is 4.4–35 V

- Input logic is available separately

- Heat sinking with package design source

- Embedded system design protection and automatic thermal shutdown

- Noise immunity

9.5 ROBOTIC ARM

The design shown in Figure 9.11 is a robotic arm. It consists of a 360-degree rotation base with mounted structure that can rotate 360 degrees as per its programme. So it is also called a programmable robotic arm. This arm can move to pick up and place the objects according to its programming. The base of this design is attached to the shoulder that supports the elbow to

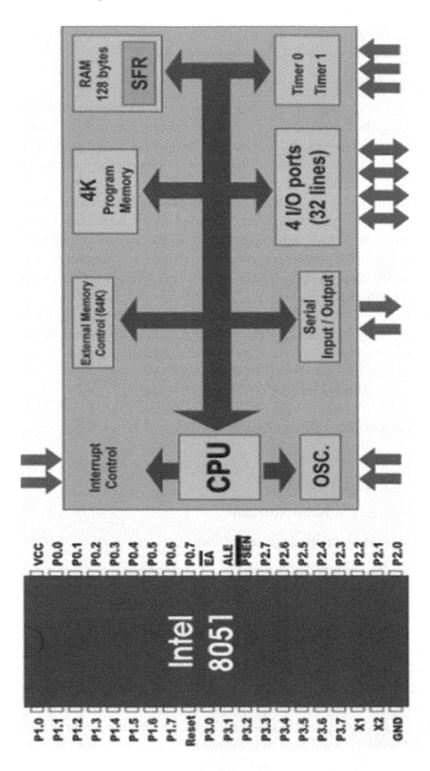

FIGURE 9.9 Driver IC.

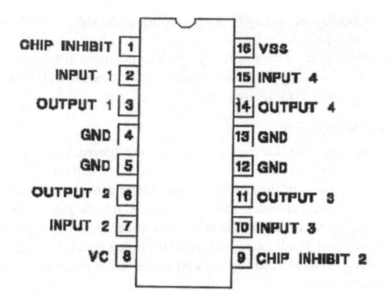

CHIP INHIBIT	1	16	VSS
INPUT 1	2	15	INPUT 4
OUTPUT 1	3	14	OUTPUT 4
GND	4	13	GND
GND	5	12	GND
OUTPUT 2	6	11	OUTPUT 3
INPUT 2	7	10	INPUT 3
VC	8	9	CHIP INHIBIT 2

FIGURE 9.10   L293 IC.

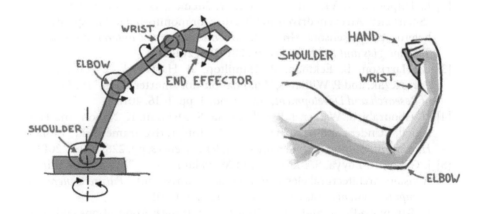

FIGURE 9.11   Robotic arm.

move up and down as per the requirement of the system. Further motion of the motor up to 180 degrees can be controlled by the wrist of the arm. The object can be rotated and picked upside down or vice-versa. Finally, a number of objects can be lifted to different places with proper grapping. The gripper controls movable jaws to grip the respective objects at various desired positions according to the Electromyogram signals generated by gestures of the hand.

Finally, all parts of the robotic arm work simultaneously to pick up and place objects from one place to another as per the movement of the gesture which is generated [13–15] by a hand.

9.6 CONCLUSION

It has been observed that the aspect of work is achieved. In the first part, EMG unit is set up. This consists of filters, amplifiers, and raw signals. The filter section converts the raw EMG signal into a convenient form that is fed into the microcontroller. Then, the driver IC interfaces with the actuating quantity for the movement of the arm as per the instructions of the EMG raw signal. Finally, manipulation of the task is realized by the microcontroller. This means that functionality of the task is programmed in the microcontroller.

REFERENCES

[1] M. A. Mazidi, *8051 Microcontroller and Embedded Systems*, 3rd Edition, Pearson Education India, 2006.

[2] L. Filipponi, A. Vitaletti, G. Landi, V. Memeo, G. Laura, and P. Pucci, "Smart city: An event driven architecture for monitoring public spaces with heterogeneous sensors," In *IEEE 4th International Conference on Sensor Technologies and Applications*, pp. 281–286, 2010.

[3] C. Harrison, B. Eckman, R. Hamilton, P. Hartswick, J. Kalagnanam Paraszczak, and P. Williams, "Foundations for smarter cities," *IBM Journal of Research and Development*, vol. 54, no. 4, pp. 1–16, 2010.

[4] H. Chourabi, S. Walker, J. R. Gil-Garcia, S. Mellouli, K. Nahon, and H. J. Scholl, "Understanding smart cities: An integrative framework," In *45th Hawaii International Conference on System Sciences*, pp. 2289–2297, 2012.

[5] J. Gubbi, R. Buyya, S. Marusic, and M. Palaniswami, "Internet of Things: A vision, architectural elements, and future directions," *Future Generation Computer Systems*, vol. 29, no. 7, pp. 1645–1660, 2013.

[6] D. Bandyopadhyay, and J. Sen, "Internet of Things: Applications and challenges in technology and standardization," *International Journal of Wireless Personal Communication*, vol. 58, no. 1, pp. 49–69, 2011.

[7] I. Ganchev, Z. Ji, and M. Droma, "A generic IoT architecture for smart cities," In *Irish Signals & Systems Conference and China-Ireland International Conference on Information and Communications Technologies*, Limerick, 2014.

[8] R. Jalali, K. El-khatib, and C. McGregor, "Smart city architecture for community level services through the Internet of Things," In *Proceeding 18th International Conference on Intelligence in Next Generation Networks*, Paris, 2015.

[9] S. Prasad, G. Singh, and A. Rai, "Artificially controlling the limb movement of robotic arm using MI with EMG sensor," *International Frequency Sensor Association*, vol. 89, pp. 39–51, March 2008.

[10] F. V. G. Tenore., A. Ramos, A. Fahmy, S. Acharya, R. Etienne-Cummings, and N. V. Thakor, "Decoding of individuated finger movements using surface electromyography," *IEEE Trans on Biomedical Engineering*, vol. 56, no. 5, pp. 1427–1434, 2009.

[11] Y. C. Du, C. H. Lin, L. Y. Shyu, and T. S. Chen, "Portable hand motion classifier for multi-channel surface electromyography recognition using grey relational analysis," *Expert Systems with Applications*, vol. 37, no. 6, pp. 4283–4291, 2010.

[12] C. Kendell, E. D. Lemaire, Y. Losier, A. Wilson, A. Chan, and B. Hudgins, "A novel approach to surface electromyography: an exploratory study of electrode-pair selection based on signal characteristics," *Journal Neuroeng, Rehabil*, vol. 9, p. 24, 2012.

[13] S. Day, *Important Factors in Surface EMG Measurement*, Bortec Biomedical Ltd Publishers, pp. 1–17, 2002.

[14] D. Staudenmann, I. Kingma, D. F. Stegeman, and J. H. Dieen, "Towards optimal multi-channel EMG electrode configurations in muscle force estimation: a high density EMG study." *Journal Electromyogram Kinesiol*, vol. 15, no. 1, pp. 1–11, 2005.

[15] M. Malboubi, F. Razzazi, and S. M. Aliyari, "Elimination of power line noise from EMG signals using an efficient adaptive laguerre filter," In *2010 International Conference on Signals and Electronic Systems (ICSES)*, pp. 49–52, 2010.

CHAPTER 10

Performance Analysis of Adaptive Cruise Control Using Arduino for Healthcare

Manish Pakhira, Rupali Shrivastava, Vishesh Kumar Mishra, and Lokesh Varshney

CONTENTS

DOI: 10.1201/9781003227595-10

10.1 INTRODUCTION

Today, safety and comfort are the major issues faced by an individual while driving a car on the road, particularly in heavy traffic. Pressing the brake, clutch, and accelerator on a repetitive basis while driving in heavy traffic or on a long journey is very painful for the body. Also, drivers have to change gears continuously which is very uncomfortable. To solve these problems, adaptive cruise control (ACC) is an ideal choice. The authors have designed a model based on ACC, an automotive feature that allows a vehicle's cruise control system to adapt vehicle speed to the environment. The system proposed is modified to make it less complex, low-cost, and flexible. In the front of vehicle, a radar system is attached to detect whether other automobiles in the ACC vehicle's path are moving slowly. If any vehicle is moving slowly, the ACC system slows down the vehicle speed and clearance is controlled. The first phase of the system is that it needs the user's current speed information, which can be taken through analogue speed controller. Next, the user needs to provide their desired speed. Then, the system needs to know the distance to the next vehicle, which can be obtained from an ultrasonic sensor attached in front of the vehicle, followed by inputs to be taken through Arduino. After analyzing inputs, Arduino processes data and gives the desired output.

The automotive world is moving toward increased safety and renewable energy [1]–[4]. Recent research involves Internet of Things (IoT) and automation. Most of the daily-life gadgets (i.e., Smart Light, Smart Watch, Automatic fan) are automated today, and the automated ones are more cost-effective and time-saving with respect to the physical ones [5]–[6]. Today, safety systems perform a significant role in protecting lives, and here the authors of [5] have tried to increase the daily-life safety by ACC, which is easy to develop and it helps to avoid the road accidents due to over speeding of vehicle.

ACC is an advance application of electronics, machine learning and OpenCv. In reality, cruise control means maintaining a selected constant speed without continuously using an accelerator pedal. But in the extended or enhanced version of cruise control, i.e., ACC, speed can vary from user to user and position to position of the vehicles. All the components the author of [6] has used to develop this module are very basic and easily available.

Any processing is much faster in microcontrollers than in computers. Arduino is a computation unit, which takes in inputs to provide alerts and precision action for electronic speed controller (ESC) and relay. It also displays the next vehicle's speed for taking the required action. ACC allows a vehicle's cruise control system to adapt the speed of the vehicle to the environment. Here, the ultrasound-sensor is being used as a radar-system. The system can detect the variable speed of the front car accurately and can give alerts if the front car is slowing down, speeding up, or is at a very close distance, which is controlled through an autonomous throttle control and applies the break as less number of times as possible [7].

This chapter presents various sections like basic components required, working of the model, and study on the different situations of rider. And the main goal is making a cost-effective, autonomous, and flexible model which helps in preventing road accidents.

10.2 BUILDING THE MODEL

For building any hardware system one needs some basic hardware and software components. Similarly, for creating this system some hardware components, as well as software components, were used. For measuring distance between the two vehicles, an ultrasonic sensor (HC-SR04) has been used. The current speed of the vehicle is obtained through an analogue speed receiver. The alert is shown through LEDs when the distance between both the vehicles is less than what is required. All these inputs are stored in the computational unit called Arduino which contains the program of the system. And there are few other components that have been used, like breadboard, connecting wire, and cardboard for giving stiffness to the model. Now, let's see how all hardware components are used in the system:

10.2.1 Ultrasonic Sensor

This is an electronic device which can measure distance of any goal by emitting ultrasonic sound waves, and converting reflected sound waves into the electronic signal. Velocity of an ultrasonic sound wave (i.e. 340 m/s) is faster than that of normal audible sound waves (i.e. 331 m/s). Ultrasonic sensors have two main components, one is a transmitter that emits ultrasonic sound waves and another is a receiver that receives the sound waves (emitted by transmitter) that hit any object. In this system, the authors have used a HS-SR04 series sensor, as shown in Figure 10.1.

Ultrasonic sensors mainly measure the time (while calculating the distance) taken between the emission of the sound (by the transmitter) and its reception (by the receiver) [8]. In this system, the ultrasonic sensor senses the distance between a user's car and cars/bikes or other obstacles

FIGURE 10.1   HC-SR04 sensor.

in its path. It transmits (by transmitter) ultrasonic sound waves every ten microseconds and if there is any object, the sound wave hits the object and reflects back to the receiver of the sensor. Then the sensor measures how much time the sound wave took to get back to the sensor. Everyone knows the relation between distance and time, i.e. in eq. (1)

$$\text{Distance}(D) = \text{Time}(T) * \text{Wave Speed}(V) \qquad (1)$$

However, for the ultrasonic sensor, as the sound wave travels the distance twice, one consequently must divide the distance (D) by 2 to get the actual distance. Now the value of distance is sent to the Arduino by the sensor. In this system, the sensor measures the distance in the period of every one second.

10.2.2 Arduino Uno

Another component used in system is the Arduino Uno board. This is a freely available microcontroller board that based on the microchip ATmega328P microcontroller and it is developed by Arduino.cc [9]. There are several sets of digital and analogue input/output (I/O) pins in this board. The board contains 14 digital I/O pins which includes six capable of PWM (pulse-width modulation) pins, and six analogue I/O pins.

It is programmable with Arduino IDE software and can be powered by USB cable or via external nine-volt battery [10]. The author of [10] used the Arduino Uno model for the system. Now, when the value comes to the Arduino, it takes the action for ESC/breaking. Thus, after every one second Arduino receives a new value of distance. The new value of distance may remain the same or may be different. Now there are three possible actions that can be taken by the Arduino:

1. If the new distance remains same the Arduino maintains the same speed of user's car and gives a signal.

2. If new distance is greater than older speed then Arduino also increases user's car speed (up to set speed) and gives a signal.

3. If new speed is less than old speed then Arduino speed down user's car's speed and gives an alert.

10.2.3 Analog Speed Receiver

Every car/bike has a maximum speed and a minimum speed. Arduino can only receive the voltage limit of 0–5 volt. So, 0 volts indicates the minimum speed of the car/bike and 5 volts represents the maximum speed of the car/bike. Hence, other speed values (i.e. in between the maximum and minimum speeds) are represented by the other voltages (i.e. in between 0–5 volts). Ergo, that's the reason of using an analogue speed receiver in the system. An analogue speed receiver is shown in Figure 10.2.

The authors have also used some (2/3) LEDs for the alert purpose. Red LEDs for speeding down/breaking, green LEDs for speeding up, and yellow LEDs for maintaining the same speed. Some other components also

FIGURE 10.2   Analog speed receiver.

like, breadboards, jumper wires, and cardboard have been used as well. Breadboards and jumper wires are used to connect all components properly and for a better connection. Cardboard is used to give strength to the whole system.

10.2.4 Arduino IDE

Arduino IDE (Arduino Integrated Development Environment) is open-source software. It is used to write and upload code easily into the board [11]. So to build the system, the author of [11] have used Arduino IDE software. Whatever code is used to build the system, everything was written on Arduino IDE software.

10.3 WORKING OF THE MODEL

In this model of ACC, the user needs to give three inputs to the system. Firstly, the system needs the user's current speed which can be taken through analogue speed controller. Secondly, the user needs to provide his desired speed which is the maximum speed that user wants but due to some traffic or obstacle can't gain. Thirdly, the system needs a distance of the next vehicle which can be obtained from an ultrasonic sensor that is attached to the front of the vehicle. The ultrasonic sensor triggers ultrasound waves from one circle (dot), and the echo (hole) receives that and, based on the time taken to receive waves, calculates the distance.

Arduino is a computation unit, which takes all these inputs to provide alerts and precision action for control of ESC and relay. It also displays the next vehicle's speed for taking the required action. After receiving these inputs, Arduino processes data and gives the desired output. It gives the speed of the next vehicle by calculating given input.

Consider that the starting distance between a user's vehicle and the next vehicle is $d1$ and after t duration of time the distance changes to $d2$. So, the change in distance after t duration of time is given by

$$\Delta d = (d2 - d1) \tag{2}$$

$\Delta d/t$ gives the difference of speed between the user's vehicle and the next vehicle. By adding this difference to the user's speed, it provides the next vehicle's speed.

As the change in the distance becomes negative, Arduino sends an alert by blinking LED light and therefore sends a signal to ESC and relay for controlling speed and braking respectively. As soon as ESC and relay

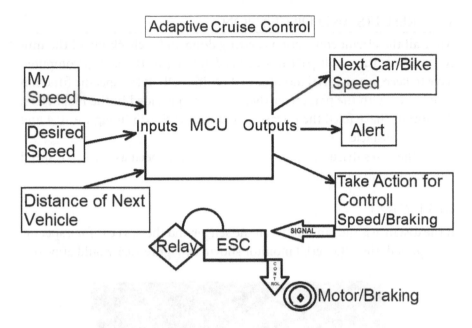

FIGURE 10.3   Block diagram of ACC.

receive the signal, the required action is completed. ESC is used to configure the speed of the user's vehicle and a relay is used for configuring the
brake of the vehicle. A block diagram of the working model is shown in
Figure 10.3.

10.4 EXPLANATION OF CIRCUIT DIAGRAM OF THE MODEL

In this model of ACC, a supply of 5 V is given to the microcontroller—
in this case Arduino. The supply is provided through a power jack. Now,
the 5 V supply is taken from Arduino and connected to the VCC pin of
the ultrasonic sensor and another GND pin of the ultrasonic sensor is
grounded. The other two pins are the echo pin and trigger pin. The echo
pin is connected to PIN 2 and the trigger pin is connected to PIN 3 of the
Arduino. The positive terminal of LEDs is connected to PIN 5 as output
and another terminal of LEDs is connected to a 220 Ohms' resistor whose
other end is perfectly grounded. Out of three pins in the relay, the GND
pin is connected to the ground, the VCC pin is connected to 5 V, and the
remaining IN pin of the relay is connected to PIN 4 of the Arduino. In
the analogue speed controller, the GND is grounded and the VCC is connected to 5 V. The data pin of the analogue speed controller is connected
to analogue pin A3 of the Arduino.

10.5 RESULTS AND DISCUSSION

After all the circuit connections getting done and rechecking all the unit conversions, the result part has come. Plugging in the laptop/computer, have to keep eye on serial monitor of Arduino IDE. If the specific function is matched with the parameter then it do the provided in the loop. Here, the authors have fixed the ultrasound sensor as shown in Figure 10.4 and move the obstacle.

All the cases discussed here consider user-car speed as the same speed and primary-distance (d1) as the constant.

10.5.1 Case 1

If the distance is the same between the sensor and the next car for a specific time period, then the serial monitor shown in Figure 10.5 would appear.

FIGURE 10.4   Ultrasound sensor on the vehicle.

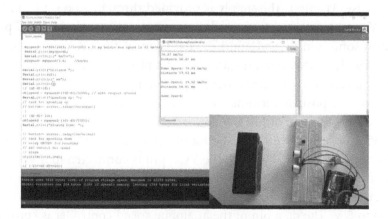

FIGURE 10.5   Screenshot of serial monitor for case 1.

10.5.2 Case 2

If distance is increasing between the sensor and next car for the specified period, the serial monitor shown in Figure 10.6 would appear.

10.5.3 Case 3

If the distance is decreasing between the sensor and the next car for the specified time period, the serial monitor shown in Figure 10.7 would appear.

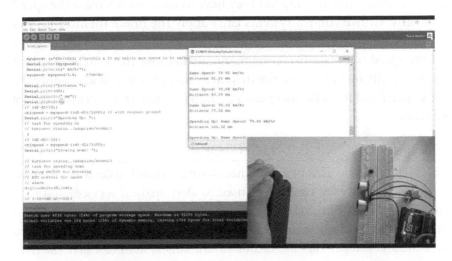

FIGURE 10.6 Screenshot of serial monitor for case 2.

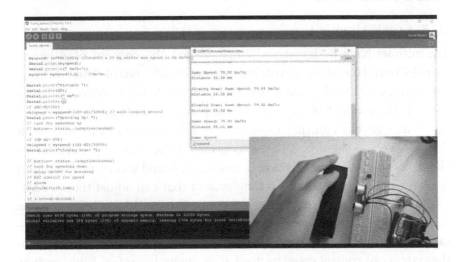

FIGURE 10.7 Screenshot of serial monitor for case 3.

The whole model was working fine, and the accuracy of the sensor was very high. At first, the authors faced an issue regarding unit conversion but later fixed it. One should follow the general purpose input output (GPIO) pins located allocated inside the code for connecting all input output devices to the microcontroller. This process is presented very simply but it was not so simple when we started out. Perfect teamwork resulted in this success. A teamwork is more effective to solve out a tricky technical problem. Cruise control users today can set a fixed speed in their car/bike while driving but they have to increase/decrease the speed manually. Cruise control systems can't apply the brake themselves. But in the case of the system developed by the authors, there is no need to set the speed (increase/decrease) manually because the system can do it automatically by measuring the distance. This system can also apply the brake itself and it is also inexpensive compared to current cruise control systems.

10.6 IMPACT ON HEALTH SECTOR

This system can be used in ambulances to make travel easier and faster for emergency services. It can also be used to alert medical services and relatives in case of any mishaps. It sends an alert message to the medical staff and guardians to apprise them of the incident so that help can be provided on time and the person's life can be saved. This system can save the lives of people who have been in an accident as the medical services can be provided as soon as possible.

10.7 CONCLUSION

The authors have designed and implemented the concept of Adaptive Cruise Control based on an Arduino microcontroller board. This Arduino-based project is an adequate method for gear-changing systems and speed control. The model has made the whole process user-friendly and easier. The system makes a journey more comfortable as the user can relax while driving because there is no need to press on the brakes, accelerator, or clutch continuously. It also makes a journey safe and secure as it's a highly sensitive speed control and braking system that can adjust the speed by itself depending upon the speed of the next vehicle. It also helps medical staff to provide services as soon as possible by sending them an alert message. The system is very simple and of low cost. And more important, as the system can control speed by itself, it also helps lower fuel consumption.

The main disadvantage of this arrangement is that it is unable to detect small-width objects or obstacles and because of not being fully automated, users have to be careful while driving. This Arduino-based model has been tested successfully and can be implemented on vehicles to avoid road accidents. The model may become more reliable if we use multiple number of ultrasound-sensors. Then, it will be more secure and collision-free to get through bends in the road.

REFERENCES

[1] Snigdha S., Lokesh V., Rajvikram M. E., Akanksha S. S. V., Aanchal S. S. V., R.K. Saket, Umashankar S. and Eklas H. "Performance Enhancement of PV System Configurations Under Partial Shading Conditions Using MS Method," IEEE Access, 2021.

[2] Snigdha S., Manasi, Meenakshi and Lokesh V. "Comprehension of Different Techniques used in Increasing Output of Photovoltaic System," Presented in International Conference on Electrical and Electronics Engineering (ICEEE 2020), SPRINGER, 28th–29th February 2020, NPTI, Faridabad.

[3] Ritu S., Rakesh Y., Snigdha and Lokesh V., "Analysis and Comparison of PV Array MPPT Techniques to Increase Output Power," Presented in International Conference on Advance Computing and Innovative Technologies in Engineering (IEEE), 4th & 5th March, 2021.

[4] Alisha K., Abhayanand K., Snigdha and Lokesh V., "Increasing the Efficiency of Solar PV Array Using Sudoku Configuration," Presented in International Conference on Advance Computing and Innovative Technologies in Engineering (IEEE), 4th & 5th March, 2021.

[5] Deepak K. R., "Arduino Based: Smart Light Control System," *International Journal of Engineering Research and General Science (IJERGS)*, Volume 4, Issue 2, March–April 2016, ISSN 2091–2730.

[6] Onyema K. N et al. "Suburbanization Case for Engineeringship as 1st Tier Government Industry Intrinsic Partner", International Journal of Engineering Research and General Science Volume 4, Issue 2, pp. 6–11, March–April, 2016. http://pnrsolution.org/Datacenter/Vol4/Issue2/109.pdf

[7] Samir K et al. "Adaptive Cruise Control, Technical note", Sep 2013. https://automobiletechinfo.blogspot.com/2013/09/adaptive-cruise-control.html

[8] N. Amin, and M. Borschbach, "Quality of Obstacle Distance Measurement Using Ultrasonic Sensor and Precision of Two Computer Vision-Based Obstacle Detection Approaches," 2015 International Conference on Smart Sensors and Systems (IC-SSS), 2015, pp. 1–6, doi: 10.1109/SMARTSENS.2015.7873595.

[9] A. A. Galadima, 2014 11th International Conference on Electronics Computer and Computation (ICECCO), Abuja, pp. 1–4.

[10] Shashidhar K., Ramakrishna G. and G.K.D Prasanna Venkatesan. "Arduino Based Monitoring and Control System for Heavy Vehicles,"

International Journal of Engineering Trends and Applications (IJETA), Volume 5, Issue 1, January–February 2018, ISSN 2393–9516, www.ijetajournal. org. Published by Eighth Sense Research Group.

[11] Fezari, Mohamed & Al Dahoud, Ali. (2018). Integrated Development Environment "IDE" For Arduino.

Index